Managing Risk in the  artment

MANAGING RISK IN THE VOLUNTEER FIRE DEPARTMENT

CONCEPTS, METHODS, AND PRACTICES

JOSEPH R. NEDDER JR.

Fire Engineering®
BOOKS & VIDEOS

Disclaimer

The recommendations, advice, descriptions, and methods in this book are presented solely for educational purposes. The author and publisher assume no liability whatsoever for any loss or damage that results from the use of any of the material in this book. Use of the material in this book is solely at the risk of the user.

Copyright © 2020 by
Fire Engineering Books & Videos
110 S. Hartford Ave., Suite 200
Tulsa, Oklahoma 74120 USA

800.752.9764
+1.918.831.9421
info@fireengineeringbooks.com
www.FireEngineeringBooks.com

Senior Vice President: Eric Schlett
Operations Manager: Holly Fournier
Sales Manager: Joshua Neal
Managing Editor: Mark Haugh
Production Manager: Tony Quinn
Developmental Editor: Chris Barton
Cover Designer: Brandon Ash
Book Designer: Robert Kern, TIPS Technical Publishing, Inc.
Cover Photo: Mark Blair

Library of Congress Cataloging-in-Publication Data Available on Request

ISBN print 978-1-593704-88-9
ISBN epub 978-1-593706-59-3

Printed in the United States of America
1 2 3 4 5   24 23 22 21 20

I would like to dedicate this book to two friends and colleagues: Chief Bobby Halton, group publisher, and Diane Rothschild, executive editor. Our fire service is blessed to have a professional leader such as Bobby—a leader who sees the light when it comes to training. Thank you for giving me a chance back in 2008 and thank you for asking me to write this book. I'm honored to have you as a friend.

Diane, you have helped me to dig deep within and find new topics to delve into. If someone had told me 20 years ago I would be writing and getting published in *Fire Engineering*, I would have thought that person was crazy. I'm a small-town firefighter and major advocate for training, but to write for a national publication, much less publish a book? Diane, it has been a delight working with you and your team. Thank you for all the encouragement. And, yes, I will write more for you, but give me a chance to catch my breath from this project!

I would like to thank Mark Haugh and Chris Barton from Fire Engineering Books & Videos. Mark, for going through the book proposal submitting process with me and patiently answering many questions. And to Chris, my editor, who has been a good and fun ally to work with. Thank you for putting up with me. With your help I think this book will make a difference in the volunteer fire service.

Finally, to Janet, my wife of 43 years. Thank you for being the understanding spouse who has always encouraged me to follow my passion, the fire service. Through thick and thin, good times and difficult times, times of sorrow and times of joy, you have always been right there by my side.

—Joe Nedder
Uxbridge, MA
April 2020

Contents

Preface   ix

1   **Managing Risk: Individual Common Sense First Steps** 1
Introduction   1
Risk, Volunteers, and Some Statistics   4
Managing Risk: Individual Common Sense First Steps   9

2   **Identify and Understand Methods to Limit and Manage Risk** 23
Risk Management   23
The Four Steps of a Risk Management System   28
The Five Situational Awareness Questions   32

3   **Managing Risk** 37
Step 2. Identify the Dangers and Risks and How They Affect Us   37
Prioritizing the Risks and Dangers   71

4   **Control, Reduce, and Eliminate Dangers** 75
Step 3. How Do You Control, Reduce, or Eliminate the Dangers
      Identified and How Can You Reduce the Risks to the
      Firefighters?   76
Survivability Profile: When an Incident Involves Trapped or Missing
      Civilians   109

5   **Managing Risk in the Volunteer Fire Department** 113
Step 4. Maintain an Ongoing Evaluation of the Incident and What Is
      Happening   114
Indicators and Warning Signs During Your Ongoing Evaluation   120

6   **Additional Ways We as Volunteers Can Reduce Risk** 129
The Incident Safety Officer (ISO)   129N
The Rapid Intervention Team (RIT)   133
Training as a Method for Risk Reduction   141

Notes   153
Index   157

Preface

Risk? What risk? I'm all set

How this book came about. I joined a small, rural New England volunteer department on April 2, 1977. It did not take long for me to realize how much I loved it! I began to realize that training was the most important thing, so I began to take training classes outside of the department along with making nearly every department drill. Eventually, I was made the department training officer, then I attended classes at the state fire academy and was recruited as a state instructor. I later moved to Uxbridge, MA, where the fire department was a combination organization, and joined as an on-call member. The move opened new doors and new friends. It was a great experience because it was a busy department and I went to a lot of fires. In 2013, realizing this was a young person's job, I reluctantly retired. Even though I am retired from on-the-line service, my love for and dedication to training continues to this day.

So how did I get here, with this book?

As a volunteer firefighter I was always involved in the *doing* part. I took thousands of hours of training and skills development because I wanted to be the best I could be. I eventually became an instructor—or teacher, as I prefer. Great teachers inspire, and I have been told by many that my greatest teaching skills, besides my knowledge, are my passion and ability to inspire. I was a department training officer, worked for the state's fire academy for over twelve years, and eventually in 2006 started my own small training group, Cross St. Associates. The more training I conducted, the more I realized that there was a tremendous need for better training for our volunteer community. Most programs were, and many still are, one-size-fits-all. We as volunteers need training programs that address and understand our needs. This does not mean watering down programs and skills, but rather finding ways to present them in a meaningful way that is not only the *how* but also the *why*.

In the beginning I focused on two areas: the basics, for they are the foundation of everything we do, and rapid intervention training (RIT). RIT has been a passion for me. I truly believe that if something goes wrong, rapid intervention might be the only thing that separates you from life or death. From the time I began teaching the skill in 2000 I have trained literally thousands of individuals in the discipline of rapid intervention and even written a textbook on the subject.

I was approached in 2008 by the Cheshire, MA, Volunteer Fire Department and asked if I could produce a Firefighter I/II training program for 40 students. My answer was yes, and our group developed and successfully conducted a

200-hour training program. One hundred percent of the students who elected to go on to take the Pro Board certification test passed.

A decade later, I continue to be dedicated to providing basic training. In 2018, the Oak Bluffs, MA, Fire Department (all volunteer) approached me about conducting a Firefighter I/II basic training class. They had a training need and the scheduled, one-size-fits-all options available on the mainland were not workable for them. You see Oak Bluffs is on the island of Martha's Vineyard and you can only get there by boat or plane, so they were really training isolated. Working with Chief John Rose and the training officer, Deputy Chief Shawn Broadly, we began to explore what they could do within their budget while still fulfilling their needs. We came up with a 170-hour program offered in a schedule that was doable and reasonable for the students. I saw this not as another job, but rather an opportunity to provide quality training to a group of volunteers who could not get what they needed—basic training—with a schedule that worked for all. The results were spectacular!

As time passed, I realized that more than just hands-on skills needed to be taught. I identified a large gap of unfulfilled training specifically for our officers. I began to address this need in 2010 by developing one-day seminars that focused on basic information all volunteer officers really need to know. I initially called it "Volunteer Company Officer Considerations." The class caught on quickly and it is now a 21-hour volunteer officer's class offered in components. Components of that class, which I call "First Due Volunteer Company Officer, the 10 Foundation Stones of Knowledge" include need-to-know subjects like fire behavior, size-up, reading smoke, building construction, decision making, and many others. As I continued to develop and expand the program, I thought it should have a risk management component as this was a hot subject, but I knew next to nothing about the topic. After some research, I began presenting this segment, but I hated it! It was boring and the concept made no sense to me and consequently the students did not respond well to the material. I quickly realized two things: (1) the presentation did not do its job, and (2) it stunk! I could either remove it from the program or really fix it to make the topic meaningful and understandable. So I began that journey of research, education, and development of a better program.

During this time I met Chief Bobby Halton, editor-in-chief of *Fire Engineering* magazine, and Executive Editor Diane Rothschild. I had some great conversations with Chief Halton, who encouraged me to write an article for the magazine. With this type of encouragement, I decided to submit a proposal to teach at the 2010 FDIC International. Thankfully, the program was approved and selected. I was on cloud nine, and can proudly say that I have continued to teach every year since. I also wrote numerous articles for Diane and the magazine, which were published in the *Volunteer's Corner*. One of the challenges of submitting proposals to present at the FDIC is to keep coming up with new topics/classes. About eight years ago I started to propose classes that were directly aimed at our volunteer service, as I believed and still do that there is not enough out there really focused on us and our needs! From that, the idea of a class "Managing Risk in the Volunteer Fire Service" came into focus. I wasn't sure it would be something the FDIC would even consider, but it kept floating in and out of my mind. A month or so later, I ran into Chief Halton at FDIC 2017.

What do you think the odds are of running into the one guy you're hoping to see among the 35,000 attendees at FDIC? Was it fate, perhaps? He was heading somewhere and only had a moment, but I took a chance and said hello, then asked, "Hey Chief, do you think the topic of managing risk in the volunteer fire service might be something the FDIC would consider for a class?" I expected something along the lines of let me think about it, but instead got, "Wow, that's a great idea!" I was urged without any promises to submit that topic. I thanked him, said good-bye, and started walking the opposite way. About five seconds later I heard, "Hey Joe, wait a minute!" I turned around to see Bobby running back to me with a big smile. Then he asked, "Joe, would you consider writing a book for us on that topic?" Wow, I was floored!

> *Now folks as most of you don't know me, I need to stop here and give you some background. A few years previously I had written a book entitled* **Rapid Intervention Crews** *for another publisher. A book like that on such an important topic was, to say the least, a big project. When it was all done and published I was very happy with the results, but I was tired and vowed to everyone I knew that I would never write another book again. It just took a lot of time, commitment, energy, and many sleepless nights! Instead, I continued to write for Diane and* **Fire Engineering**. *It was and is fun, and Diane kept encouraging me to write more!*

After Bobby asked the question, I hesitated about two seconds and said "Sure!" So much for my vow of never writing another book again!

This book has been an interesting project. I have learned much from my research and now realize how lucky I was as an active, on-the-line firefighter. I took risks all the time. Did I realize it? Probably not, but even if I had, I wanted to be in the middle of all the action. Sound familiar? I am now a disciple of risk management and on a mission to help spread the word to those in the volunteer fire service that this is a very important, need-to-know subject.

1

Managing Risk: Individual Common Sense First Steps

Introduction

This book is for all members of the volunteer fire service: from the newest recruit to the most experienced chief officer or 40-year veteran. In it we address traditional risk management, but we also talk about all the other aspects of risk we expose ourselves to on a daily basis and how we can reduce or eliminate those risks. Risk management can be a complicated and intimidating subject. For this book I avoid discussion of community risk assessment and other aspects of the prescribed risk management models. A review and discussion of all aspects of risk management would cause this book to be more than double the size. I wanted to make this book easy to read and to understand. My ultimate goal is to get you to buy into the risk management concepts presented here, realizing that they can and do make a difference. We focus on responding to fires and other emergencies. I believe that once we in the volunteer service have a better handle and working model on reducing risk to ourselves, we can look to expand our scope.

To begin with, let's look at who we are, as volunteers, in the big scheme of things. According to the National Fire Protection Association (NFPA), volunteers represent about 70% of all the firefighters in the United States. It is in communities with populations of 10,000 or less that the vast majority of volunteers are found. Regardless of the size of the community to protect, we must be ready and able to do our job. Part of that job is protecting ourselves. Volunteers fight the same dangers as those who are full-time firefighters protecting their communities. Fire does not discriminate and neither does risk. To me what this means is that if I want to be a firefighter, I must be a firefighter. We must learn to do our jobs well and limit risks by doing the right thing and working smarter. Many

1

times while teaching, students ask me why we need to know this stuff because "we are only volunteers!" In my opinion, a statement like that is a put-down and insult to our noble service. We are not "only" anything. We are firefighters. This is, of course, if we have the skills and knowledge to do our jobs safely and effectively. I have learned that there are two types of people on the fireground: firefighters and civilians dressed as firefighters. I would hope that on every fireground there are only firefighters, but if not, then use this book to help fix things.

Risk management. Risk/benefit analysis. Situational awareness. Risk a little to save a little. Risk a lot to save a lot. Had enough? I joined the volunteer service in 1977 and actively served for almost 37 years. I continue to teach and provide resources to the fire service, much of which is focused on us volunteers. During much of that time, like so many of us, I really did not understand what the words "risk management" meant, why they were important, and frankly had little interest in spending any time on the subject. I wanted to fight fires, not worry about risk! However, as time progressed, the nature and science of fire got more intense. I got wiser and began to think about the risks that I was taking and how they could affect me. But even then it took many, many years to actually understand the problems, danger, and liabilities I was exposed to. I started to read more about risk, began going to the annual *Fire Engineering's* FDIC (Fire Department Instructors Conference), and attended many classes, trying to broaden my knowledge and be the best I could be.

In these 40+ years I have seen, participated in, or heard of many actions or lack of actions that have been either very close calls or caused injury or death. In general, very little risk management was applied. In reading reports from the National Institute for Occupational Safety and Health (NIOSH), in discussion with many other volunteers, and in applying my life lessons, it has become quite apparent that the vast majority do not fully understand risk management, what our risks are, or how we can reduce risk to our members. The reasons vary, however I feel the top reasons are *complacency* and *complexity*. Complacency has been a plague for many. It is internally bred, intentionally or unintentionally, and many members have no idea about what dangers they truly face. The fact that "they never had a problem" is used to block something new, and any call for reason or acceptance of the increased dangers we face are ignored or shouted down. Add to this complexity. So many pick up an article or book on risk management and find themselves in a sea of intellectualism. For many this is not a bad thing, but for others the theories and concepts make no sense or lack application. Many see risk management as something for the city departments or larger jobs. I think the subject needs to be presented in basic and simple terms. I'm not saying we are not capable of understanding complexity or intellectualism, but let's face it, most of us want the facts presented in an easy and useful manner, and in a way that will be easy to present to our membership to get buy-in!

Managing risk is one of the most important yet frequently ignored fireground management skills for the volunteer service. It is further compounded in our volunteer fire service by a lack of understanding and training in a risk management system; why we need it, how it works, and how to use it. How often have we seen firefighters taking risks that are not necessary? Yes, firefighting is a risky and dangerous job, but it should be a calculated risk! Volunteers put a lot on the line every time they respond, whether it be to a structure fire, brush fire, or vehicle accident. This book discusses what is risk today. It explores why and how it is different from the risks of 20 years ago. It challenges you to accept that the risks have never been higher, and that volunteers are being injured or dying every year. The book presents a different risk management process and shows specific actions that we as volunteers can take to reduce and manage the risks we are exposed to. We also examine other risk management techniques not thought of in the traditional sense.

In this text we review NIOSH statistics, contributing factors, and key recommendations to help us all see the problem. The statistics are based upon volunteers only (which includes on-call departments) and cover the years 2006 to 2017. In that period of time, NIOSH investigated a total of 1,195 line of duty deaths (LODDs), reporting that volunteers accounted for 527 (or 44%) of those deaths! The information and statistics I reference are from this baseline of 527 volunteer LODDs.

In addition to the NIOSH data, I have received some incredible help from Rita Fahy, Ph.D., the manager for applied research at NFPA, to gather some very interesting statistics and present them in this book. Again, it is all for volunteers only (on-call included). As you look at both the NFPA and NIOSH statistics combined, they tell a very solid story.

My goal is to give you useful and applicable thoughts, systems, and ideas that can and will reduce risks on firegrounds. As you read this book, pause and contemplate situations that have happened, or, if you were lucky, almost happened. Then compare those experiences with what is presented here and use that knowledge for future calls to make a difference.

In this book several points and ideas are frequently repeated. I do this to make a learning point. You need to hear, see, and read something more than once in order for it to sink in and remember it. Some of the most important themes I repeat over and over are "we are not all the same," "we are different," "one size does not fit all," and, a personal favorite I always seem to hear, "that's not the way we do it here."

Much of what you can learn from this book comes down to one thing: *training*. In Chapter 6 we discuss training. I hope that the need for it and changes that might be required in your department's training will be very much evident to you.

My friends, this book is for us!

So let's begin.

Risk, Volunteers, and Some Statistics

What is *risk*? Simply put, it is something that can cause you or others injury or death.

Volunteers put a lot on the line each time they enter a burning building; respond to a brush, grass, or woodlands fire; or respond to many of the other calls they have!

Risk management is not a management issue or responsibility. Every firefighter is responsible to themselves, their brother and sister firefighters, and to their family to reduce and manage risk. We are all responsible to each other to do the right thing and to minimize the dangers to ourselves and others.

To help us better see and understand risk, let's begin with looking at some statistics from both NIOSH and NFPA. They might only be numbers, but they represent the truth as the numbers show the facts without any filtering. It is a good starting-off point to help better define the need for risk management.

NIOSH statistics

If we look at the NIOSH LODD statistics from 2006 to 2017[1], we find that 527 volunteer firefighters died in the line of duty. If we break down these numbers we find some interesting facts.

Using specific categories listed in NIOSH reports that relate to actual interior firefighting, we find that in that time frame, 62 volunteers died in fire suppression activities. The statistics are as follows:

- Advancing hose lines—45 deaths or 8.5% of total LODDs (note that the category does not distinguish from advancing in the interior or the exterior)
- Search and rescue—13 deaths or 2.4% of total LODDs
- Ventilation—4 deaths or less than 1% of volunteer LODDs

Totaled together, we have 62 deaths or 11.7% of the total 527 LODDs that occurred in what are considered activities in or on the structure on fire. Again, not all of these deaths occurred within burning structures; the statistics do not break it down. Interestingly enough, in 2017 there were a total of 93 LODDs for all firefighters, both career and volunteer, and of that total only one was within a burning structure.

NFPA statistics

First off, let's look at some of the department profile statistics based upon the NFPA study titled "U.S. Fire Department Profile 2017."[2] It looks at all the departments in the U.S. and provides the following information:

- There are approximately 1,562,200 firefighters in the United States.
- Of that number, 682,600 or 65% are classified as volunteer firefighters, meaning both volunteer and on-call firefighters who are active in firefighting duties.
- Of the 682,600 volunteers, about half are members of small rural departments that protect 2,500 or fewer citizens.

The study also identified 29,819 fire departments in the following categories:

- Career—2,651 departments (9%)
- Volunteer—19,762 departments (65%)
- Mostly career—1,893 departments (8%)
- Mostly volunteer—5,421 departments (18%)

The NFPA statistics are slightly different from NIOSH, *but closely similar*. From NFPA I have learned that there is a higher share of volunteer injuries on the fireground as compared to our brother and sister career firefighters. We are 51.6% more likely to receive injuries on the fireground than all firefighters combined (42.1%).

The slight differences are attributed to the fact that NIOSH, as a U.S. government agency, does not investigate all firefighter deaths but rather uses a decision flow chart to prioritize investigations. (More information on this decision-making process can be found on the website https://www.cdc.gov/niosh/fire/abouttheprogram/abouttheprogram.html.)

From the NFPA research report "U.S. Volunteer Firefighter Injuries 2012–2014,"[3] published in February 2016, a higher share of volunteers incur injuries on the fireground as compared to our brother and sister career firefighters. Of the total volunteer firefighter injuries, 51.6% occur on the fireground. For all firefighters, career and volunteer combined, injuries on the fireground represent 42.1%. Volunteers also have a higher rate of injury responding/returning and at training (fig. 1–1).

If we take an honest look at the statistics and remember that not all departments are as well-equipped and progressive, the NFPA report states that the trend might likely come down to inadequate training, equipment, and the condition of the equipment.[4] I agree and furthermore will go on to say that another cause of this might be the "we are only volunteers" syndrome that causes people to believe emergency scenes are different for us!

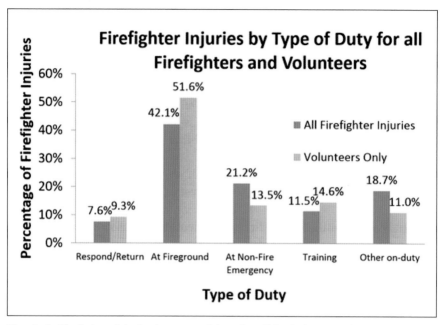

Fig. 1–1. Firefighter injuries by type of duty for all firefighters and volunteers. (Courtesy of NFPA)

Now let's look closely at figure 1–2: On-duty firefighter deaths by type of duty and year, for volunteers only. When I asked the NFPA for their help in assembling some statistics it was because I wanted to look at and share three key categories associated with volunteer LODDs: type of duty, cause of fatal injury, and nature of fatal injury. This analysis is based upon looking at ten years of volunteer deaths, from 2008 to 2017. Again, I would like to personally thank Rita F. Fahy, Ph.D., manager of applied research at NFPA, for her kind and phenomenal assistance in gathering this information.

Of the types of duty involved, the analysis shows us that the majority of deaths occurred when responding to or returning from alarms, which accounts for 34.6% of our LODDs. The second highest number of LODDs happened on the fireground (26.4%). What really bothers me when I look at the third highest number illustrated in the chart is that we are also killing ourselves in training! Training accounts for almost 11% of our deaths.

Now let's look at figure 1–3: On-duty firefighter deaths by cause of fatal injury and year (by "on duty" they mean the volunteer had responded).

Here we see that the top cause of our deaths while on duty is overexertion, stress, medical, and it accounts for 53.3% of our deaths! I have to wonder if this points to fitness, health, and the welfare of us as individuals because it sure looks

 **RESEARCH**

On-Duty Firefighter Deaths by Type of Duty and Year
Volunteer Firefighters Only
2008–2017

Year	Fireground	Responding to or Returning from Alarms	Non-Fire Emergency	Training	Other On-Duty	Total
2008	12	27	8	6	8	61
2009	12	13	10	3	3	41
2010	9	18	5	8	5	45
2011	13	10	5	3	4	35
2012	8	16	2	3	1	30
2013	17	13	5	4	2	41
2014	10	10	6	3	5	34
2015	8	11	7	1	5	32
2016	7	10	6	7	9	39
2017	7	7	10	4	4	32
Total	103 (26.4%)	135 (34.6%)	64 (16.4%)	42 (10.8%)	46 (11.8%)	390 (100%)

Fig. 1–2. On-duty firefighter deaths by type of duty and year, volunteer firefighters only, 2008–2017.

Source: National Fire Protection Association, Quincy MA, September 2018. NFPA Index# 2870

like it! Our health has become a very important topic today. The National Volunteer Fire Council (NVFC) publication *Volunteer Fire Service Culture: Essential Strategies for Success*[5] points out that in the past, health and wellness for volunteers was not considered important. The fire service has changed, the seriousness and dangers of the calls has changed, the stress and physical demands have significantly increased, and so must our attitudes and culture towards health, wellness, and fitness. We need to understand and accept this. Our activities are not for all, yet there are many other places where we can use help that is not strenuous or a cause of overexertion.

The number two cause of volunteer firefighter deaths was due to vehicle crashes (18.5%). We discuss this cause of death later in this chapter, but suffice it to say that many of these tragic crashes might be avoidable.

 **RESEARCH**

On-Duty Firefighter Deaths by Cause of Fatal Injury and Year
Volunteer Firefighters Only
2008 – 2017

Year	Overexertion, Stress, Medical	Crashes	Exposure to*	Falls	Caught/ Trapped	Struck by Object	Assault**	Total
2008	30	15	1	3	6	6	0	61
2009	25	5	1	1	4	4	1	41
2010	27	9	2	2	1	4	0	45
2011	18	4	0	4	5	4	0	35
2012	14	7	0	1	3	3	2	30
2013	16	7	1	1	12	3	1	41
2014	22	7	0	1	1	3	0	34
2015	20	3	0	1	2	6	0	32
2016	20	10	0	3	2	2	2	39
2017	16	5	0	0	1	10	0	32
Total	208 (53.3%)	72 (18.5%)	5 (1.3%)	17 (4.4%)	37 (9.5%)	45 (11.5%)	6 (1.5%)	390 (100%)

Fig. 1–3. On-duty firefighter deaths by cause of fatal injury and year, volunteer firefighters only, 2008 – 2017.

* Exposure to included: 2008–heat, 2009–electricity, 2010–fumes in manhole, bacteria, 2013– electricity
** Assault included: 2009–gunshot, 2012–gunshot (2), 2013–physical assault, 2016–gunshot (2)
Source: National Fire Protection Association, Quincy MA, 02169, September 2018. NFPA Index# 2871

Figure 1–4 examines and presents the facts and statistics of on-duty death by nature of fatal injury and year. As you look at the statistics, stop and think about the number one nature of our deaths and that it is sudden cardiac death. It accounts for 48.2% of our fatalities. That is almost half! I'm sure there is a direct correlation of the sudden cardiac deaths with the overexertion, stress, medical cause. The second highest nature of our deaths is internal trauma, crushing, and fractures. These account for 35.9%. Combined, these two natures of fatal injury represent 84.1% of our deaths.

From these tables we should be seeing that the vast majority of our LODDs are not from heroic acts of courage such as getting trapped or lost during a rescue, but rather in a category where we can and should have better control by our actions, attitudes, practices, and updating our culture.

 **RESEARCH**

On-Duty Firefighter Deaths by Nature of Fatal Injury and Year
Volunteer Firefighters Only
2008–2017

Year	Sudden Cardiac Death	Internal trauma/ crushing/ fracture	Burns	Smoke Inhalation and Asphyxiation	Stroke/ Embolism/ Aneurysm	Gunshot	Drowning	Other*	Total
2008	27	24	2	3	3	0	0	2	61
2009	22	10	0	3	4	1	0	1	41
2010	23	15	0	1	3	0	1	2	45
2011	17	14	2	0	1	0	1	0	35
2012	12	10	0	3	1	2	1	1	30
2013	14	20	3	1	2	0	0	1	41
2014	22	10	0	1	1	0	0	0	34
2015	17	9	2	1	3	0	0	0	32
2016	19	13	0	2	2	2	1	0	39
2017	15	15	1	0	1	0	0	0	32
Total	188 (48.2%)	140 (35.9%)	10 (2.6%)	15 (3.8%)	21 (5.4%)	5 (1.3%)	4 (1.0%)	7 (1.8%)	390 (100%)

Fig. 1–4. On-duty firefighter deaths by nature of fatal injury and year, volunteer firefighters only, 2008–2017.

* Other included: 2008–septic shock, heat stroke, 2009–electrocution, 2010–bleeding, septic shock, 2012–heat stroke, 2013–electrocution

Source: National Fire Protection Association, Quincy MA, September 2018. NFPA Index# 2869

Managing Risk: Individual Common Sense First Steps

In this section we explore risk in a different, nontraditional way. We discuss what our first priority should be as it concerns risk and then look at some very basic, common sense actions we can take to immediately reduce our exposure to risk.

Risk and our family

Our first priority in life should be our family. Let's stop and think about something we traditionally do not want to think about. We know that risk is something that can cause death or injury. Think about how often we have seen firefighters take risks that are not necessary. The risks we take should be calculated risks. What if something goes wrong and you die or are severely injured? What would happen to your family? Does your community have an insurance policy that would give you or your survivors benefits consistent with what you earn in your full-time job? For most of us the answer is no. How will your family survive? Many families rely on two incomes. Your death or an injury that prohibits you from working will have a negative impact on your family's financial position! What about your children? Where will the money come from to get them things they need or the things we want them to have growing up? Health insurance is getting more and more costly, and will not get any cheaper. Our families and survivors may have to depend on the generosity of the community. Perhaps your community might feel bad initially and help pay for the funeral, but then what? Who will cover all the normal expenses of life?

In the early 2000s I attended the funeral for a volunteer firefighter. He was 29, with two small children and a pregnant wife. I stood at attention on a bitterly cold winter morning, directly across from the church, watching his coffin being removed from the back of the hearse. I saw one thing only: his wife, supported by family and able to button just one button on her winter coat because of her pregnancy. It was heart wrenching. The following day she gave birth to their third child. This image and its message are forever imprinted in my mind.

I share this story because the town didn't do the right thing. There was no life insurance policy; after all, he was only a volunteer. People from across the country donated to an established family fund and donations were generous, but not enough to support the family as the children grew. The family asked the town for an annuity that would provide health insurance for them until the children were adults. At a town meeting (local form of government used in New England) the town officials and the townspeople voted them down. Do you really think your community will be different? I don't think so! People vote for their pocketbooks and not necessarily for the right thing. In this case, the following year the state's governor had a bill to assist the family brought forth that passed, thankfully. Can we all count on that?

The U.S. federal government has a program entitled the Public Safety Officers' Benefits Program (PSOB).[6] It provides a one-time death benefit to eligible survivors of public safety officers (us as firefighters) if the death was the result of injuries sustained in the line of duty. At the time of this writing the benefit is $343,589.00. What additional benefits does your state, city, town, or community have to assist those you leave behind? Are there survivor benefits? An insurance policy? Something? It is important that each and every one of us know this information. If there is nothing, we as volunteers need to speak as one voice and lobby, push, and demand that our loved ones are protected.

Beyond the monetary aspects, risk impacts your family in so many ways. Do you have children or grandchildren? Think of the loss they will have in their lives.

For the little ones you won't be at the elementary Christmas holiday shows, you won't see them participate in sports, you won't be there to console them when they fall and get a scraped knee. No more holidays and birthday parties to see the joy in their faces. As they get older you won't see them graduate high school, learn to drive, or take the next steps in life attending college, joining the military, or beginning a vocation! You will leave them with a large hole in their lives and in their hearts. Was the risk calculated and necessary? Is the risk worth losing your life in the line of duty, or is it preventable or manageable? Your actions and decisions affect not just you but many other people, starting with your family. Understand and remember that you must learn to manage your risk!

Reducing risks to ourselves

> **Of the 527 volunteer line of duty deaths as reported on by NIOSH, 120 or 23% were while responding.**

Common sense: responding. We need to stop and think about how many have died responding to a call, either in a department vehicle or a personal vehicle.

As volunteers we are constantly at risk in all our fire department actions. When your pager goes off you are called to action for an emergency situation, someone calling for help! What risks are you exposed to? From the outset, what condition are you in? Are you impaired by drugs or alcohol? In this day and age, we are finally facing the fact that this is a problem across the entire fire service. Had a few? Then eliminate your risk by listening on the scanner next to your recliner! Good to go? Then jump into your personal vehicle and respond! What do you know of the law in your state as it regards you responding? If you are not sure, then use common sense. Years ago, I saw a recruit firefighter who lived across the street in a classic example of lack of common sense. It was snowing and had been for most of the day. The roads were snow covered and slippery. My pager went off for a possible chimney fire and as I went out to my truck, I saw the recruit pulling out of his driveway slipping and sliding all over the place. I proceeded, in four-wheel drive, at a cautious pace based upon the road conditions. About a mile down the road I saw the classic furrows in the snow indicating a vehicle had gone off the road. I pulled over and, you guessed it, it was the recruit! I called out to see if he was okay. Fortunately all that was bruised was his ego and he'd already called for help, so I proceeded to the station and my call to duty without him. A few weeks later with similar weather conditions, except a lot more snow, the pager opened for a possible structure fire. As I went to my truck, I saw my youthful neighbor repeat his adrenaline-fueled driving antics as he fishtailed out of his driveway. Almost like déjà vu as I responded, *at the identical place on the road*, I once again saw those telltale, car off the road furrows in the snow! I stopped and wasn't surprised to find our recruit had not learned his lesson. This time he was out of his car and trying to run through the snow to catch a ride from me. I called out to see if he was okay and then promptly left!

Why did I leave him? Two reasons. The first was he was exhibiting a lack of common sense and driving without due regard for his safety and the safety of others, responders included, who might be on the road. The second reason was a person who exhibits behavior such as this is in my opinion prone to freelance or take an action that could endanger other firefighters at the scene. I know some will think I abandoned a brother in need! To the contrary—I abandoned a person who we cannot count on to do the safe thing. Perhaps I saved someone that night from injury or worse.

Copyright © Robert Gaughran

When we drive our personal vehicle or apparatus as if possessed by crazed demons, with total disregard for anyone else on the road, we are asking for trouble. Putting 732 blinking lights on your car or truck does not give you the authority to drive like an idiot. Nor do the lights, sirens, and air horns on the rigs! Sometimes I think that today's road rage was invented by volunteers many years ago, when some little old lady driving a 1910 Ford Model T would not yield the road. Slow down! Drive with due regard for all and drive defensively. Believe me, you will get there.

NIOSH Statistics

Of the 527 volunteer line of duty deaths, 120 or 23% were while responding.

Think of the number! We lost 120 volunteers before they even got to the emergency scene. If we look at this category and the 120 deaths a little closer and isolate the cause of injury as vehicle collision, we find the following information about the ages of the volunteers:

Under 21 = 12.5%
21 to 30 = 13%
31 to 40 = 11%
41 to 50 = 16%
51 to 60 = 19%
61 to 70 = 17%

The data also shows that in the under 21 to 40 age group, 82% died of trauma caused by the crash. For those in the 51 to 80 age group, the cause of death was heart attack 62% of the time.

NIOSH reports always list what they term contributing factors and key recommendations to prevent another similar death. In the LODD reports for vehicle collisions we find some factors and recommendations repeated over and over. They include the following:

- There was no department- or state-required driver training program.
- Fire departments should ensure that all persons responding in emergency apparatus are wearing and restrained securely by seat belts at all times while the vehicle is in motion.
- Fire departments should provide training to driver/operators.
- Fire departments should ensure that all fire service vehicles are operated safely, taking into consideration the type of emergency and route of travel to the scene.
- Departments should provide training to driver/operators as often as necessary to meet the requirements of NFPA 1451, and incorporate specifics on rollover prevention into their standard operating procedures (SOPs).
- Departments should develop and enforce SOPs that require the mandatory use of seat belts in all moving vehicles.

I would like to add that seat belts should be used by all, at all times, when responding in the apparatus or in a personal vehicle. This is common sense! In many of the reports of the 120 deaths while responding, the firefighter was ejected or partially ejected. This clearly indicates no seat belt. In today's day and age, if I asked to take your small child for a ride in my car and was not going to secure them with a seat belt, none of you would let the child in my car! We recognize that seat belts are important and needed, so why do so many think that they are not needed while responding? Wear your seat belt and, if you are the officer in the apparatus, do not let the vehicle move until all your firefighters are securely belted! Want to save lives? Want to reduce risk? Want to reduce needless injury and death? Then step up and do the right thing—no excuses, no compromises. Seat belts really do save lives, including ours, so use them!

If you are an apparatus driver/operator, you need to understand that a personal vehicle is typically in the 3,500-pound range and pickups are more like 5,000 to 6,000 pounds. Now you climb into a tanker (tender to appease those who had to change the terminology after almost 100 years!) that typically weighs 60,000 to 65,000 pounds GVW, an engine at about 47,000 pounds, or an aerial that is 75,000 to 85,000 pounds. How these fire service vehicles operate, handle, and stop is completely different from what you drive every day! This sounds like common sense, right? But we all know that overaggressive driving of apparatus is very commonplace in the fire service. It used to be mostly tanker/tenders that we read about being involved in crashes and rollovers with injury and death. Now is seems to be spreading out to engines and ladders.

The way you drive both your personal vehicle and fire department vehicles when responding can put you, other responding firefighters, and civilians at high risk for injury or death. Slow down! Buckle up! Manage the risk by eliminating it. Learn to operate those emergency vehicles safely and you will reduce risk to all and help us all go home safe.

In the past few years it has become more and more evident that we are at extreme high risk when operating on roads and highways. In fact, in the first two weeks of 2020 seven roadway responders have been struck and killed by vehicles while assisting the public. Seven responders lost in fourteen days. The list includes fire, law enforcement, towing, and recovery responders.

They were all out there helping someone who needed their service and while they were there, a driver who was not paying attention took their lives. These numbers add to the total of 44 responders (fire, law enforcement, towing, and recovery responders) killed last year. Those statistics don't take into consideration debilitating injuries, amputations and career ending incidents that bring pain and suffering.

Responding to roadway incidents is (or has become) the most dangerous place in the fire service.

These incidents are not confined to metro departments. Many occur in rural America where volunteers are often on the roadway longer waiting for other assets to reach the scene. Studies show that the longer you are on the roadway the greater the chance for a second incident that involves you.

Because there are usually more road and highway responses than working structural alarms, departments must consider managing the risk firefighters face with the same urgency as managing the risks associated with a working fire. To assist us in getting the proper knowledge and training there is an organization called Respondersafety.com. On their website they state that part of its mission is to improve the safety of the nation's emergency responders by:

- Engaging in and promoting activities that include developing educational material to support responder safety training
- Promoting the National Unified Goal (NUG) for Traffic Incident Management (TIM) including responder safety; safe, quick clearance; and interoperable communications
- Encouraging the development of TIM Teams
- Promoting collaboration, communication and cooperation among the nation's emergency responders
- Keeping emergency responders up to date on national rules, regulations and trends related to safe roadway incident operations.

Preventing being struck is not high-tech. Situational awareness, common sense, and wearing high visibility garments will substantially reduce the risk.

Getting the training to be safer on the roadway is a snap. The Responder Safety.com website offers great online-tested training at no cost to anyone. Produced by firefighters for firefighters, the information will improve the chances that you and your volunteer fire department won't be a statistic so everyone goes home.

I highly recommend you go online and broaden your knowledge and your departments safety by taking advantage of all the free training available to you!

Common sense: You, your gear, and cancer. What else puts us as individuals at risk? *Our turnout gear!*

You have responded safely, with due regard to the law and those others operating on the road (civilians and others responding). Time to gear up! So, what is the condition of your turnout gear? When was the last time it was washed? If I examined the various components would I find that everything is intact, including a properly installed drag rescue device (DRD)? Do you wash all your gear after every fire or call that involves carcinogens, or are you trying to make it look *salty*? When was the last time they were properly washed? What about your hood, gloves, and helmet?

One of the ways we are exposed to cancer is through our personal protective equipment (PPE). It has been proven that after a fire, our turnout gear is loaded with particulates and carcinogens that can cause cancer. This applies whether you are a career or volunteer firefighter. However, in the volunteer service we have specific and unique issues that differ from our career counterparts. Many of us have carried or still carry our turnout gear in our personal vehicles. We now know that the gear after a fire is basically particulate contaminated. Primarily, it is contaminated with carbon particles from incomplete combustion. These carbon particles absorb and retain different fire gases, and this is what makes it very dangerous. These particles are what contains the carcinogens, and the carcinogens are what will harm you and possibly your family.

Our exposure is not just to our gear but to our personal clothing we are wearing under the turnout gear. Carcinogens and other dangerous toxins such as hydrogen cyanide can and do penetrate through to your skin! Contaminated gear stored in the back of your car or truck, when heated by the sun through the vehicle's windows, will actually begin to off-gas the toxins and carcinogens on the gear.

In the years I served actively, I always carried my gear in my personal vehicle. For thirty of those years this was in the back of a full-size SUV, where the storage area was a part of and fully open to the passenger compartment. For most of those thirty-seven years, I rarely washed any of my gear! Yes, I wanted it to have that look of an experienced firefighter. The downside of that is that for over thirty years my family was also exposed to all those carcinogens! Who knew? I never considered it a danger in any way. After a fire, the gear went right back into the truck and so did my family! I honestly can tell you that today I worry about any harm I have done to my family.

So, what can we do? First off, applying common sense can reduce cancer risk. Today we are fortunate with all the cancer awareness happening in the fire service. Pay attention and stay informed! Immediately stop carrying dirty, contaminated gear in your personal vehicle or the apparatus! If you work a fire and must transport your gear back, put it into a heavy-duty trash liner, seal it tight, and immediately transport it to the station for decontamination and cleaning. Dispose of the trash bag as you would any other contaminated bag. You can also use a plastic container with a lid that seals, but make sure the container is properly cleaned and decontaminated right away. Another option new on the market is bags designed to carry and isolate our contaminated gear. They allow you to bring the gear in for washing in a safe manner and the bag can be washed with the gear to be used again, vs. another plastic trash bag in the landfill (fig. 1–5). Using your gear bag to transport contaminated gear is not a good idea or safe practice. The reason is that the zipper gives you a false sense of safety and the bag is really not designed to contain the particulates and contamination during transport.

Fig. 1–5. Reusable bag for contaminated gear.

If you do carry contaminated gear in your vehicle—contained or not—back to the station for cleaning, decon your vehicle in the area the gear was transported. As it regards our rigs, we need to start thinking about removing our gear and placing it bags or containers before placing it in the rigs, and perhaps putting those bags or containers up on the hose bed to isolate them for transport. This is all new territory for so many of us, so be extra cautious and err on the side of safety!

There are other places we can look for immediate help and actions. To start with, the Firefighter Cancer Support Network is a fine and very supportive group. In 2013 they published a white paper titled "Taking Action Against Cancer in the Fire Service."[7] In that document they identify and target eleven life-saving actions for each and every one of us to take immediately, which I paraphrase slightly here:

1. Wear SCBA through all stages of the fire, including overhaul.
2. Remove as much of the bulk contamination as possible while still at the fire scene by performing gross decontamination.
3. Wipe soot from your head, neck, jaw, throat, underarms, and hands using wet wipes immediately after the fire.
4. Change and wash your clothing right after leaving the fireground. (Remember, the carcinogens have penetrated your PPE and are on your clothing.)

5. Shower after the fire. (This should be done right away, especially before returning to work or family to reduce exposure to the contaminates to others.)
6. Ensure that all your gear is properly cleaned right after the fire.
7. Do not transport or take contaminated clothing home or store it in your vehicle. (A suggestion has been made that placing the gear in a zippered duffel or gear bag, or placing it in a large tub with a lid that seals, will provide some protection before the gear can be properly washed.)
8. Decontaminate the interior of the apparatus after the fire.
9. Keep your gear out of living and sleeping areas and limit your family's exposure. (During bad weather or extreme cold have you ever brought your dirty bunker pants into the house to keep them warm and speed up your response? I did.)

The last two actions are not related to gear but to individual choices:

10. Don't use tobacco products.
11. Use sunscreen.

In addition, the National Volunteer Fire Council publishes a poster, available for the asking, that lists 11 best practices for preventing firefighter cancer (see fig. 1–6).

Both of these organizations are excellent resources and available to you. Take advantage of the materials and educational materials they offer. Reach out, contact, and explore. The issue of cancer in our fire service is real. Take the time to learn about it and do something about it. Both the NVFC and the FCSN have help lines you can reach out to if you find a cancer diagnosis in your future. You are not alone; these people are there to help you. The brother and sisterhood of the fire service extends far beyond what we think it is. As a firefighter, it is very important to know that there is always someone there for you no matter how bad things are!

As this book was in editing, a new organization was brought to my attention and it is important to share with you all, it is the "15-40 Connection." They began by focusing on early cancer detection for all, but quickly realized that as firefighters, our risk is higher. Their website is https://www.15-40.org/get-involved/fire-service/.

To best explain what they are all about, I quote from their website:

> 15-40 Connection is different from other cancer organizations and shares two new ways to think and talk about cancer:

BEST PRACTICES
for Preventing
FIREFIGHTER
CANCER

1 Full protective equipment (PPE) must be worn throughout the entire incident, including SCBA during salvage and overhaul.

2 A second hood should be provided to all entry-certified personnel in the department.

3 Following exit from the IDLH, and while still on air, you should begin immediate gross decon of PPE using soap water and a brush, if weather conditions allow. PPE should then be placed into a sealed plastic bag and placed in an exterior compartment of the rig, or if responding in POVs, placed in a large storage tote, thus keeping the off-gassing PPE away from passengers and self.

4 After completion of gross decon procedures as discussed above, and while still on scene, the exposed areas of the body (neck, face, arms and hands) should be wiped off immediately using wipes, which must be carried on all apparatus. Use the wipes to remove as much soot as possible from head, neck, jaw, throat, underarms and hands immediately.

5 Change your clothes and wash them after exposure to products of combustion or other contaminants. Do this as soon as possible and/or isolate in a trash bag until washing is available.

6 Shower as soon as possible after being exposed to products of combustion or other contaminants. "Shower within the Hour"

7 PPE, especially turnout pants, must be prohibited in areas outside the apparatus floor (i.e. kitchen, sleeping areas, etc.) and never in the household.

8 Wipes, or soap and water, should also be used to decontaminate and clean apparatus seats, SCBA and interior crew area regularly, especially after incidents where personnel were exposed to products of combustion.

9 Get an annual physical, as early detection is the key to survival. The NVFC outlines several options at **www.nvfc.org**. "A Healthcare Provider's Guide to Firefighter Physicals" can be downloaded from **www.iafc.org/healthRoadmap**.

10 Tobacco products of any variety, including dip and e-cigarettes should never be used at anytime on or off duty.

11 Fully document ALL fire or chemical exposures on incident reports and personal exposure reports.

FUNDING PROVIDED BY

Fig. 1–6. This National Volunteer Fire Council poster lists best practices for cancer prevention.

15-40 Connection is focused on educating and empowering people about early cancer detection.

This education helps individuals become aware of the early warning signs of cancer. Most cancer organizations focus on research for a cure, treatment or support.

There are also many cancer organizations that focus on prevention. Unfortunately, we still don't know what causes all cancers, so while some preventative measures can reduce risk; it can't remove the risk completely. Research shows that detecting cancer early improves effectiveness of cancer treatment and also improves the chance of survival, which is why 15-40 Connection is empowering individuals to be aware of the early warning signs to give them their best chance at effective treatment and survival.

15-40 Connection aims to educate and empower individuals with the skills to recognize subtle health changes in themselves, rather than rely only on medical professionals.

Through 15-40 Connection's "3 Steps Detect," individuals learn how to become active participants in their own health care so cancer as well as other illnesses can be diagnosed earlier. The result is a quicker return to health and most importantly lives saved.

The risk factor of occupational cancer is a critical issue facing the nation's firefighters. Members of the fire service have a 9% higher risk of developing cancer and a 14% greater risk of dying from cancer than the general public. Fire departments across New England, and the largest municipal fire department in the United States, the New York City Fire Department, all agree:15-40 Connection's 3 Steps Detect is a vitally important "missing piece" needed to help reduce cancer mortality in the fire service.

Detecting cancer early can be the difference between life and death.

The 15-40 Connection differs from most other organizations in that they focus on teaching us the importance of *early detection*. I would encourage each and every one of us to look into this most noble of endeavors and see what they can offer you and your organization!

While writing the draft for this book, the United States government passed H.R.931: Firefighter Cancer Registry Act of 2018.[8] It requires the Secretary of Health and Human Services to develop a voluntary registry to collect data on cancer incidence among all firefighters, career and volunteer. It is a beginning and finally an acknowledgement that firefighters are affected by cancer! In addition the NFPA is beginning to track cancer deaths.

In wrapping up this section on cancer and our health we need to stop and digest what we present here and ask what does it mean to us volunteers? First, remember that we are all in a higher risk group to get cancer. Secondly, we can reduce our risk of cancer by taking simple, common sense actions. We can all benefit from this! Let's follow these easy and basic guidelines to limit and mitigate expose to things that can adversely affect our health. You, your family, and all firefighters will be better off.

This chapter is all about learning to managing risk by using some individual, common sense first steps. In the following chapters we look at what risks we repeat over and over, examine a four-step risk management system, and discuss other core foundations of the volunteer fire service that can help to reduce risk to all.

2

Identify and Understand Methods to Limit and Manage Risk

In Chapter 1, we discussed how risk affects not only us as individuals but also other members on our team, and, equally important, our families. When managing risk, using common sense can and will reduce hazards. In this chapter we look at risk from the perspective of the incident commander (IC) and outline the first step of a risk management system: Develop situational awareness of the incident.

Remember that the first due officer or acting company officer is the initial IC. It's the job and cannot be avoided. In fact, the initial IC may end up in charge of the entire incident, regardless of rank.

Risk Management

So, what exactly is *risk management*?

According to *NFPA 1250: Recommended Practice in Fire and Emergency Service Organization Risk Management*,[1] risk is "a measure of the probability and severity of adverse effects that result from an exposure to a hazard." Risk assessment is "an assessment (action or an instance of making a judgment about something) of the likelihood, vulnerability, and magnitude of incidents that could result from exposures to hazards."[2] Risk control is "the management of risk through stopping losses via exposure avoidance, prevention of loss (addressing frequency) and reduction of loss (addressing severity), segregation of exposures, and contractual transfer techniques."[3] Finally, risk management is "the process of planning, organizing, directing, and controlling the resources and activities of an organization in order to minimize detrimental (adverse, unpleasant) effects on that organization."[4]

When reading these definitions, we may feel the topic is overly complicated and understood only by intellectuals. It is this complexity and intellectualism that has contributed to our lack of understanding and actual participation in risk management.

Want further proof? Reread the NFPA definition of risk control: "The management of risk through stopping losses via exposure avoidance, prevention of loss (addressing frequency) and reduction of loss (addressing severity), segregation of exposures, and contractual transfer techniques."

I was an on the line firefighter for over 36 years, and in all the training, seminars, and conferences I attended and participated in, I was never taught about a contractual transfer technique and I bet you weren't either.

The standards are important, but we are obligated to find ways to bring it to our membership in more easily understood terms. I'm not saying that we are not intelligent enough to understand these definitions, but let's face it, understanding and using this type of information is a lot easier when it is presented in plain English.

Let's try again from a different perspective: what exactly is *risk management*? If we look up these words individually in a dictionary, *risk* is something that presents the possibility of loss (death) or injury, and *management* is the act of managing, to control.

Now we can define risk management as the act of trying to control things that can cause us injury or death. We must learn to manage or control risk at all levels of our organization, from the department chief to the newest recruit. We all have a stake in and obligation to manage risk.

Please note the NFPA is an important organization that fulfills an important role in the fire service, but I think I know our members and what makes sense to them. Our first obligation is to ourselves and our department. To fulfill that obligation, we need to ensure our members understand what we are talking about and buy in to the idea. The key is simplicity. From a manager or chief officer's position, it is important to not only understand the topic, but to work toward understanding the more complex aspects of risk management. As managers, our job is to take the number and complexity of ideas, information, or thoughts and simplify them for our team, as executing these ideas is critical to everyone's job.

Understanding risk and how we can limit and manage it

Risk is a part of a firefighter's job. It's what we signed on for; however, none of us signed up to be seriously injured or die. We all have a tremendous ethical responsibility to each other to prevent the worst from happening. On the fireground, we look to the person in charge—the incident commander—to protect us. We are relying on that person's knowledge, skills, experience, and common sense. A competent IC should consider the controls needed to reduce the risk to which the members operating at an incident are exposed. This also carries over to the company officers. Together, they must manage the risks we are exposed to while understanding the consequences of certain predictable actions that are routine to the job.

Predictable actions and their consequences

In the fire service we take certain predictable actions, almost without thinking, such as advancing the attack line, opening up on the fire to knock it down and extinguish it, raising ladders for rescue and egress, and venting. Yet we seldom stop and think about the consequences of these actions. It's almost like we are on automatic control; get off the rig and do what we are expected to do. These predictable actions all have consequences, some of which are good and some of which are bad. We tend to only remember the good and think that not doing the action is bad. Well, given the type of extreme fire conditions we fight today, we must stop and think about what are we doing and what could go wrong with such an action.

One example of a predictable action is the basic task of horizontal ventilation. We have, in the past, been taught ad nauseum that we need to ventilate a fire quickly for a variety of reasons, such as lifting the smoke so we can crawl under it, or improving the interior environment for any trapped civilian and to help expose the fire location. Terms like "coordinated fire attack" were hammered into us from the beginning of our training.

Today's fires are very different than the fires old timers like me fought. Many of us might have relatives or friends who were volunteers during the '60s, '70s, and early '80s. When I started firefighting we wore cotton duck coats; tin, plastic, or leather helmets; pull up boots; and work gloves from the local hardware store. Not everyone wore a SCBA and we all removed them during overhaul. We rode the rear step regardless of the weather; in fact, many rigs were open top cabs (as in no roof). Obviously, we have progressed greatly from that time. So why are so many still holding onto out-of-date tactics?

Many of these old timers have no idea or concept of how fires today are much more dangerous. These increased dangers come from the fact that the fuel load of today (where most everything is made of petroleum-based products/plastics) burn hotter and faster.[5] It is a fact that today's residential structure fires burn significantly faster and hotter than 30 to 40 years ago based upon modern construction techniques and synthetic contents (most everything in our homes is made of petroleum based products, plastics). Homes back then were not as weather tight, there were no thermal pane windows, or materials that eliminated air infiltration like exterior house wrap or spray foam insulation. Things are very different today and we must learn and understand this!

Today we can arrive on scene and find a "vent limited" fire. The nature of the construction, insulation, and tight windows and doors deprives additional oxygen from entering the building and feeding the fire if all openings are shut. If you don't understand this, or "buy in," you might order venting and opening up, sending a company into the structure on a fire attack. They are advancing into what is now called the flow-path. The *flow-path* defined "is the route that air/oxygen takes when it heads in to feed the fire and the route that fire, smoke or heat heads for, to leave the building."[6] To us when we "open up" we need to think of it as a bidirectional flow, the hot gases and smoke are flowing above and out while fresh, oxygen enriched air is flowing in below feeding the fire with oxygen. Today we need to know and remember that within the exiting smoke is fuel (the gases). When the fire flashes, so will this fuel-enriched smoke, and if a company is advancing in the flow-path they are now in a very bad place.

Another consequence we must understand is backdraft. We've been taught that under 16% oxygen levels, fires cannot sustain a flame so they smolder. In today's tightly sealed home, fire will deplete the oxygen supply and, if it gets under the 16% range, will cause the fire to reduce to a smoldering stage while still generating high heat.[7] The fire slows down in growth as it starves for oxygen. This is a perfect set up for a backdraft. Then we start venting by breaking glass or opening doors, as we have been taught, introducing a sudden supply of fresh, oxygen-enriched air, and we have most likely created a backdraft! These are violent and can and have hurt firefighters.

Things were different in the old days, but things have changed and so must we. The tactics we were have been taught all these years have not, until recently, changed. Basically, we have been fighting hotter, more aggressive, and dangerous fires with tactics based upon what fires were like in the '50s, '60s, '70s, and '80s. It's time our tactics caught up! In Chapter 3 we discuss this in much more depth.

The predictable act of venting a window or opening a door is a lot more complex today. When we order the companies to open up or vent we need to remember that the sudden introduction of fresh, oxygen-rich air can cause the fire to grow to a flashover event very quickly. The companies pushing in are being exposed to an immediate danger that can burn or kill them. Today, the basic act of ventilation is something that needs to be thought through as a predictable act, and must be coordinated and controlled by the incident commander. The incident commander must control the risk of this predictable, routine action by identifying the flow-path, and coordinating and controlling the venting. As simple as all this sounds, it is something that not everyone understands. We are still hearing about firefighters getting trapped, burned, injured, or worse when caught in flashover or near flashover situations.

Another example of a predictable, routine action that is a risk is assigning a tactic (such as fire attack) to a company incapable of performing the task due to a lack of training or physical capabilities. The assignment might be made while the crew is en route and the IC is not aware of the makeup of the company. Making assignments is part of our routine actions, but giving the assignment to an incapable company might lead to consequences such as injury or death. To control the risk you should either not assign the tactic to that crew, or, once the crews' makeup is discovered, reassign the tactic to a more capable company. A few pages back, I quoted NFPA's definition of risk control. It specifically states "contractual transfer techniques." So what does that mean in plain English? It means if the IC assigns (makes a contract in legal terms) a tactic to a company and then realizes they are incapable of it, the commander needs to control the risk by reassigning (transferring) the tactic to a company capable of completing the task.

Actions have predictable or expected results, and knowing these predictable results can save firefighters from injury or worse. As ICs, whether titled or assumed, we need to understand these actions and consequences in order to be safe and effective for the safety and survival of our firefighters.

Risk management and the incident commander

Let's look at risk management from the perspectives of the incident commander and, in some respects, the company officer.

Perhaps the most common term we hear as it regards risk is the *risk-benefit analysis*. This is not just a fire service analysis. It can be found readily in any number of professions and applications, including business and medical. The basic premise of a risk-benefit analysis is to ask, what are the risks if we perform a certain action, how likely are the risks to happen, what are the benefits to taking the risk, and how likely are we to get the outcome (benefit) we hope for?

Back in the 1980s Chief Alan Brunacini taught us the following:[8]

- We will risk our lives a lot, in a highly calculated and controlled manner, to protect a savable human life.
- We will risk our lives a little, in a highly calculated and controlled manner, to protect savable property.
- We will not risk our lives at all to protect lives or property that is already lost.

I remember reading about this in Fire Engineering magazine. It sounded really good, but as a small-town volunteer firefighter, I confess I thought that it did not apply to us. After all, Chief Brunacini was a large city chief. Things are different in small communities, right? Little did I know.

Over time, this has pared down to the popular saying, "Risk a lot to save a lot, risk a little to save a little, and risk nothing to save nothing." I have always interpreted it to mean the following:

- Risk a lot to save a lot means to save a life.
- Risk a little to save a little means the victim is dead.
- Risk a little to save property not worth saving means the property burning is beyond saving and no risk to our lives should be taken.

For a risk-benefit analysis we ask, is the risk worth the gain? All this sounds simple, but the other question that needs to be asked is does every officer really have the skills, knowledge, and capabilities to make these critical types of decisions? Many years ago, when attending a RIT (rapid intervention team) class, I was taught the saying, "You might be able to say it, but if you can't show it, you don't know it." The point is, we can all say the words but can we truly perform if and when needed?

The Four Steps of a Risk Management System

To better prepare for these types of decisions let's look at the components of a risk management system. The beginning of this chapter outlines some terms from *NFPA 1250: Recommended Practice in Fire and Emergency Service Organization Risk Management.* These terms are important to know and understand; however, it boils down to what I present as a basic, simple, and easy-to-understand definition: Risk management is the act of trying to control things that can cause us injury or death.

A risk management system is a tool to use in both pre-incident planning and on-scene decisions to evaluate and reduce firefighters' exposure to injury, loss, or death. There are many different interpretations of the system. Some are verbatim from the *NFPA 1250* standard. Here, I have chosen something that meets the standard but is easy to understand and use (fig. 2–1).

The four steps of this risk management system are as follows:

Step 1. Develop situational awareness of the incident.
Step 2. Identify the dangers and risks and how they affect us.
Step 3. Think about how to control or eliminate the dangers and risks identified and how to reduce the risks to the firefighters.
Step 4. Maintain an ongoing evaluation of what is happening.

Fig. 2–1. Four-Step Risk Management System

Step 1.
Develop situational awareness (SA) of the incident

Situational awareness is a term most of us have heard. Basically, SA means being aware of what is happening around us. Many different industries use SA including medical, aviation, the military, law enforcement, and (of course) firefighting.

Wikipedia defines situational awareness as "Situational Awareness involves being aware of what is happening in the vicinity to understand how information, events, and one's own actions will impact goals and objectives, both immediately and in the near future. One with an adept sense of situation awareness generally has a high degree of knowledge with respect to inputs and outputs of a system, an innate 'feel' for situations, people, and events that play out because of variables the subject can control. Lacking or inadequate situation awareness has been identified as one of the primary factors in accidents attributed to human error. Thus, situation awareness is especially important in work environments where the information flow can be quite high and poor decisions may lead to serious consequences (such as piloting an airplane, functioning as a soldier, or treating critically ill or injured patients)."

If we try to simplify this definition and apply it to the fire service operating at an emergency scene, it means that we need to be constantly aware of what is going on around us and understand how the situation in front of us and our actions will affect the goals that have been set. The definition goes on to talk about a high degree of knowledge. It also points out that a lack of situational awareness has been a primary factor in accidents and deaths attributed to human error.

For the IC, this means constantly observing the overall situation when operating at an emergency scene. Company officers (and frankly all firefighters) are expected to also maintain a continual situational awareness regarding the company's assignment, the progress the company is making, the conditions and environment in which they are operating, and how have the conditions changed or are changing.

> **All firefighters need to remember the basic premise of firefighting: Conditions are always changing—either they get better or they get worse.**

As the IC, take a look around and evaluate the situation. Many years ago, long retired Framingham, MA, Deputy Chief Jack Corcoran (deceased) taught me something I will never forget. When first arriving, look at the emergency situation (fire, accident, brush fire, HAZMAT, etc.) and ask three questions:

1. What do I have?
2. Where is it going?
3. What do I have to do to stop (or mitigate) it?

This is the initial size-up sequence. Simple, to the point, and an excellent beginning, it follows along very nicely with the four-step system I outline above. These questions can help us to begin and maintain situational awareness.

Asking, what do I have? is a quick mental evaluation of the incident. This is the beginning of the size-up and it helps in creating situational awareness, which should be ongoing. This first question will begin to help in making decisions. Figure 2–2 guides us to answer our SA and first-due questions.

On arrival, our initial size-up tells us we have a two-story home with an attached garage fully involved and the fire most likely extending horizontally into the first and second floors and the attic. Vehicles in the driveway indicate there might be people trapped, or they may have escaped.

Fig. 2–2. Viewing this scene tells firefighters the situation and what questions need to be answered. (Photo courtesy of Mark Blair.)

Look at what is on fire and where is it located. Which way will the fire spread? Are there civilians trapped or unaccounted for? Think about what initial actions will contain the fire or make for an effective rescue. I have often said that as volunteers, most of us are well acquainted with the initial situation of a little help, a little water, and a lot of waiting for both! For those blessed with large memberships and great municipal water supplies, it might seem odd to have to wait for help and water; however, for all too many volunteer fire departments in the United States, this statement rings true.

So, understanding this basic premise means we need to start stretching the correct lines (size and flow) for extinguishment. Fire attack is a basic action for us, but, we need to pause and ask five additional questions. These questions are part of the situational awareness survey. While asking or analyzing them, remember that there are variables that differ on every call!

The Five Situational Awareness Questions

1. What is the current on-scene staffing level?

Many if not most volunteer organizations are used to working with a lot less staff than city and urban firefighters, but there are things that need to be done and done safely. Say for example the first-due apparatus is staffed by four but the driver is only that—a driver, two firefighters are fairly new with little or minimal training, and the fourth firefighter is qualified and capable. (This lack of training might be due to age, physical health, or department policies.) What do we really have for a company? In reality we have a pump operator and a crew of one to start the fire attack. We all know that sending in one firefighter on a fire attack is not going to happen.

2. Is more help needed?

Once we know the answer of what do we have, the next question should be, is more help needed? The above situation indicates that a lot more help is needed. Order it now. Don't wait to see if anyone else will show up because that is wasting valuable time. More help is always needed. Help can come from additional members of the department or from mutual aid. Many departments have predetermined run cards that list what mutual aid company is to be called and when. Many departments create these run cards based upon the number of alarms, and typically range from 1 to 10. The incident commander radios to the dispatcher for a second alarm, and the dispatcher can just look at the list and know who to call. A run card reduces radio traffic and makes life a lot easier when we have a raging fire and don't have to think who to tell the dispatcher to call.

If the department still operates with the IC telling the dispatcher what departments to call for mutual aid, then before calling, he or she should think of specific equipment that may be needed such as a pumper or ladder. For example, a request for mutual aid puts an engine on the road. Shortly thereafter, the IC also requests a ladder from them. Now they are sending two pieces, but the ladder is what is needed first. Perhaps a few extra seconds of thought might have helped to identify the more urgent need for the ladder and thus have it respond with the staffing first. While assessing staffing and the need for additional help, consider added variables such as the type of incident and the weather. Having preplanned run cards for mutual aid greatly assist in the IC not having to spend valuable time selecting the departments wanted, and help in getting them there quicker.

Different types of incidents can greatly increase staffing needs. To begin with, if we are in a rural water area, we know that we are going to need a lot of help at the tanker (tender) fill site and dump sites. A large fire with a good start will most likely require a lot of manpower as companies rotate. We also know that weather can greatly affect the need for more help. Extreme heat, cold, rain, or snow wears people down. We need to be thinking of the members' health and safety. In these situations it is very wise and in the interest of safety to get more help right away.

Some volunteer departments are in more suburban areas with enough staffing and quick responding mutual aid departments. Even if this is the case, the IC still needs to make a quick evaluation of the staffing level. Do the skills of those initially on scene match the needs of the incident? Is it one of those days when a lot of people are out of town or attending a function and cannot respond? Consider the size and scope of the incident and think about where it is going. We have been taught that fire will double in size every thirty seconds to two minutes (depending on what book or article you read). It is a non-debatable fact that fire in a structure, given adequate oxygen (well vented) and depending on what fuel is available to burn, will quickly increase in size and quickly spread throughout the structure. What this means to us is that the present fire has the potential to increase in size significantly if the needed help and resources are not on scene now, or soon, to contain it. We need to anticipate the help that will be needed and call for it early in the incident.

3. Are the members on scene capable with the emergency before you?

When assessing staffing we need to honestly ask if the members are capable. Look at who is available and ready, consider the tactics/tasks that need to be done, and then determine if they are capable. Do they have a capable officer? All volunteers are not equal. Some states have different levels of firefighters. For example, New York state has a designation for Interior Firefighter. The designation is earned by completing training to a certain level. But even with that, we need to question and know an individual's capabilities.

I use myself as an example. I have always considered myself a good and capable firefighter. I had the skills, knowledge, and physical capability to do my job. After I hit 60, though, I felt my body starting to resist some of the more strenuous firefighting activities. I was starting to feel very exhausted after calls. My chief and I had a long talk we decided that I would focus on being an apparatus operator, driving, and running the pump or ladder. A few years later, we had two large structure fires within a five-day period that required me to do more than be an operator. The calls were in the middle of a very warm summer and frankly, the heat kicked the "you know what" out of me. This is when I realized that even though I was in good health, as per my doctor, being a firefighter was a young person's game and I was no longer young. So I told the chief what had happened, that I did not want to collapse and die on the fireground, and had decided to retire. Could I have fulfilled other roles like running rehab or filling bottles? Sure, but that was not what I wanted to do. There comes a time in all our lives where we need to accept the obvious and step aside. I share this personal story because we have to make a decision if the firefighters standing before us are ready and capable.

Consider what members can do, their health, and their physical abilities. This is not discrimination but practicing safety for all involved. It is our job to manage risk and do what we feel and know is in the best interest of all the members operating on scene.

I would also say, and to some this might be controversial, this also applies to being a company officer. A company officer is responsible for executing tactical assignments from the IC. These include fire attack/suppression, ventilation, and search. This means we are in or on the building, pulling hose, climbing ladders, crawling, and performing physically. I personally cannot understand how people who are company officers in their 60s and 70s think that it is okay to retain the rank and let another firefighter (no rank) fulfill the role of leading a company into a burning building. Sorry if this is upsetting to read, however, I am writing about risk, not politeness or department politics. When we cannot fulfill our job on the line with the company, we need to honestly evaluate our position. Perhaps it is time to step down and allow someone else to step up. Valuable knowledge and expertise can still be shared in training, or perhaps as an aid to the IC. How often do we hear "It's a young person's job"?

> **Rank is authority, not power. It is also leadership.**
> **A good leader knows when it is time to let others step up.**

When we consider whether more help is needed and look at who is available, consider the task at hand and their training. Are these members trained and capable of performing the task? Is the entire company trained and capable? Attending a training session and not participating is not the kind of training to keep skills sharp. Remember all those nights of a poor membership turnout for training? Honestly ask, are these people capable of doing what needs to be done and get out safe and sound? Remember, poor training or a lack of skills shows very quickly.

4. What equipment do we need and what is currently on scene?

An initial size-up helps to identify what additional equipment is needed. Is the department rural, in an area that does not have hydrants and always relies on tankers (tenders)? When we call mutual aid from another volunteer department, it takes time to get a crew to their station and responding to the scene. Waiting or delaying the call for tankers has seen many a structure find the cellar hole. Perhaps it's time to rethink the run cards and mutual aid agreements and have tankers responding with the initial alarm. No fire? No problem, turn them around.

What other equipment is needed? Pumpers for a tanker fill site or for relay pumping? Aerial devices? If the scene is a vehicle crash, is additional equipment needed to deal with the incident? If we even think we might need help or equipment, call for it now, don't delay. It is easier to say, "Thanks, but we are all set" rather than, "How close are you? We need you now!"

5. Are you trained and able to manage the incident?

We must assess our personal capabilities. What training and experience do we have to handle the emergency? Not all calls are routine bread and butter. There is a lot more to this job than the old adage of "Put the wet stuff on the red stuff." Is it a HAZMAT situation and you have very little knowledge of how to deal with this incident? Are you new at being an IC and facing the largest fire the community has ever seen? Is the incident beyond your capabilities?

The honest answers to these five questions as they regard the initial and ongoing size-up and SA survey help in beginning the process of mitigating the emergency while taking into consideration the safety of those operating, and help to reduce the risks they are exposed to.

The skill levels of our officers in the volunteer service varies greatly. This is due to many factors including the lack of availability of quality officer training, the department's culture, and the individual's attitude towards training.

Managing Risk

Step 2. Identify the Dangers and Risks and How They Affect Us

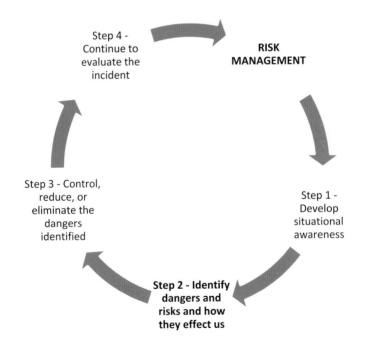

Step 4 - Continue to evaluate the incident

RISK MANAGEMENT

Step 1 - Develop situational awareness

Step 3 - Control, reduce, or eliminate the dangers identified

Step 2 - Identify dangers and risks and how they effect us

While maintaining continual situational awareness, we now need to begin to identify the dangers and risks and how they can affect us. To assist, there are twelve questions to prioritize and answer:

1. Is there an incident action plan (IAP) with identified strategies and tactics?

2. What is the fire situation (basic fire behavior)?

3. Is the flow-path identified (modern fire behavior)?

4. Is it a vent limited fire?

5. What is the smoke indicating?

6. What is the building construction type?

7. Are civilians trapped or missing?

8. What is our ability to communicate on the fireground?

9. Are we using the NIMS/ICS system?

10. What are the fireground safety considerations? (For example, is there an RIT and Incident Safety Officer?)

11. Is weather a factor?

12. Is water supply an issue?

Let's look at each of these individually. As we do, please keep in mind that each of these topics could result in a long discussion by itself. What is presented here gives an overview and enough knowledge to realize there is a lot more to learn or refresh on. There is always a lot more to learn to stay current and be ready. It is up to each person to pursue more knowledge and education.

As previously mentioned while preparing to write this book I conducted quite a bit of research on the OSHA website as it regards Line of Duty Deaths (LODDs). As part of that research I compiled statistics on not only how we are dying, etc., as discussed in Chapter 1, but also what NIOSH calls contributing factors and key recommendations. For each of the twelve questions we will explore I also include the NIOSH line of duty report contributing factors and key recommendations that I found were frequently listed, specifically in volunteer firefighter deaths from 2007 to 2017 that I researched individually and gathered facts and created statistics.[1]

1. Is there an incident action plan (IAP) with identified strategies and tactics?

NIOSH line of duty report contributing factors as they regard an IAP:

- Fireground and suppression activities not coordinated
- Ineffective ventilation
- Inadequate size-up
- Uncoordinated fire attack
- Ineffective fireground communication in regard to the incident action plan

NIOSH key recommendations:

- Ensure that properly coordinated ventilation is conducted on structure fires.
- Ensure that officers know how to evaluate risk versus gain and perform a thorough scene size-up before initiating interior strategies and tactics.

So, what exactly is an incident action plan (IAP)? It is the starting point to assess the risks and dangers. Begin with a good size-up and then continue in an ongoing manner until the scene is cleared. From our size-up we must gather information, determine the priorities, and then formulate a plan of action to develop sound strategies and tactics. In 1999, the National Fire Academy ran a program called "Multiple Company Tactical Operations" (MCTO). There were a lot of good lessons learned from that class but the one that sticks with me (and that I teach and share to this day) is their system of decision making.

Decision making on the fireground is the job of the IC. You cannot avoid it. If you don't want to make decisions, then reconsider being a company officer. The issue most of us have when it comes to decision making is that we have not been properly trained on how to do so. As a young firefighter, and then as a company officer and fire service instructor, I was afforded the opportunities to be educated by some of the best and brightest. The lessons they taught me were sometimes difficult to absorb, but once I was able to understand these lessons in basic terms, they made a lot of sense. As the first due officer, you pull up to a structure fire and you are the initial IC and perhaps the IC for the duration. The decisions you begin to make affect how the incident goes and the amount of risk that the members are exposed to. MCTO taught me that we must use a logical thought process. We know that there is always a call for immediate action. It might be from the property owners, the spectators, or your members, but everyone is waiting for you to do something. But before you can do that something, you must begin to develop your logical thought process. Let's look at the three-step logical thought process I learned from the NFA's MCTO course.[2] The three steps are: size-up, planning, and implementation.

Step 1 Size-up = Problem Identification. The first step is to begin to identify the problems before you. This is the thinking stage. It is your initial size-up, and it must continue throughout the incident. The situational awareness section of this book recommends that to assist you in your size-up, ask yourself, what do I have? and where is it going?

Step 2 Strategies and Tactics = Action Planning. The second step is to take the problems you have identified and formulate your strategies and tactics. This is the planning stage. You are asking question three of the initial size-up sequence: What do I have to do to stop it? This is the beginning of your IAP. Base your strategies and tactics on what you see and know. Don't over-complicate things. *Basic strategies* are what you want to achieve. If the size-up shows that the house is on fire (that's one of the problems identified), then a basic strategy is to extinguish the fire! A *tactic* is how are you going to do it. In this case, it is determining the need for fire attack. Obviously, there is a lot more to this. For example, if we have possible people trapped, the strategy is to rescue the people and the tactic is a primary search. This is strategies and tactics at its most basic level. Starting here and gaining experience helps this step become natural to you.

Step 3 Implementation = Tactics/Tasks. The third step is taking your determined tactics and assigning them to companies as tactical assignments. It is the acting stage. Give assignments, as much as possible, as tactical assignments. A *tactical assignment* is what you want done, a *task assignment* is how to do it. For example, tactical assignments might be, "Engine 1, fire attack, 2nd floor" or "Ladder 2, vertical ventilation." It is the company officer's job, once given the tactical assignment, to break it down into a task for the company. When the IC gives assignments as tasks they add a lot of extra radio transmissions that are unnecessary. (Remember the complaints we get about out-of-control radio traffic?) Most of the time, a simple tactical assignment is all that is needed, yet we still hear so many assignments being given as tasks. The reasons vary, but I believe the task is everyone's comfort zone. Get beyond it, think tactically. It's your job.

Now there are going to be times and conditions that require an assignment to be given as a task. It might be because you have more information than the company officer and need the action to be done a specific way. But remember, if you give an assignment as a task, you own it! If the company finds it is not working or can't be accomplished, they are going to radio and ask you what to do.

Finally, when you give an assignment as a tactical assignment you are saying that you know the crew members are capable and ready to do the job, you trust them. In our volunteer service this is a very important statement. None of us like to be micromanaged or told exactly how to do our jobs. If you don't trust them, then we have to ask is it because of their training, or is it your inability to perform as an incident commander who has little confidence in his members?

2. What is the fire situation (basic fire behavior)?

NIOSH line of duty report contributing factors regarding fire behavior:

- Lack of training on fire dynamics

NIOSH key recommendations:

- Fire departments should ensure that all firefighters are trained in and recognize the importance of situational awareness.
- Fire departments should integrate current fire behavior research findings developed by the National Institute of Standards and Technology (NIST) and Underwriters Laboratories (UL) into operational procedures by developing or updating standard operating procedures, conducting live fire training, and revising fireground tactics.

The question of "What is the fire situation?" is very easy to ask, but knowing how to answer it is not easy and requires an excellent understanding of basic fire behavior. When I talk about basic fire behavior, I'm talking about the science of it that we learned back when we read or studied our Firefighter I/II textbooks. For me, that was over 40 years ago! I remember not understanding things like "convection" and "1,700 to 1." I remember asking senior members what they meant and was told to just remember what the book said for the exam so I would get it correct, and after that don't worry about it. Time, teaching, and talking with many firefighters has shown that I was not the only one who got that advice. In fact, that advice still seems to be out there. Now, when I teach fire behavior during officer training I always ask the same question: After you first read the chapter on fire behavior or learned about it (typically during your basic training), how many times have you re-read the chapter or had an in-depth department training session on fire behavior? Statistically (gathered informally by class response), less than 20% of the students attending my classes answered that they reviewed this knowledge after first learning it. This is not a condemnation of those people or departments, but rather a call for action.

First off, why does this happen in our fire service? I believe it has a lot to do with the following:

- We never really understood the science of fire behavior and the importance of it.
- Departments downplayed it as they did not understand it.
- The focus has always been hands-on skills, so we never thought about reviewing and retaining fire behavior knowledge.
- Our culture has always been, "Just put the fire out and don't worry about the science behind it." What I call the "just put the wet stuff on the red stuff" syndrome.

Today we are seeing a lot of the same type of cultural behavior when we discuss things like modern fire science, flow-paths, and vent limited fires. If you don't understand it or are afraid of change, then ridicule it or shout it down! That attitude of "I don't buy in" seems to always want to rear its ugly head with change and new knowledge. Fire is, by nature, a scientific phenomenon/chemical reaction. We need to understand the basics of fire science to be better prepared, more effective, and, most importantly, safe.

Let's talk some basics and start at the very beginning. What are the three things we need for a fire? Oxygen, fuel, and heat (this is the fire triangle). By removing any of them, the fire will go out. Typically, as firefighters, we use water to cool the fire, thus it goes out. Using a CO extinguisher will eliminate the oxygen at an electrical fire, and turning the valve off on a burning stove or propane grill will remove the fuel. When we learn about this triangle it is most often presented as part of the fire tetrahedron, which is composed of oxygen, fuel, heat, and an uninhibited chemical chain reaction. All the textbooks show the same graphic and talk about the chemical chain reaction (fig. 3–1).

From my years of teaching experience, I have found that most firefighters do not understand this graphic and are confused. As no one wants to ever admit they don't understand something, we just ignore it and don't see any reason to ask any questions. Maybe we rely on the overly simple adage, just put the wet stuff on the red stuff.

So, what is the tetrahedron and what is it telling us? Think of a two-dimensional triangle and then transform it into a three-dimensional pyramid with three sides and a bottom. Each of the three sides is labeled either oxygen, fuel, or heat. Now take that pyramid and turn it over to see the bottom. On the bottom is the label "chemical chain reaction" (fig. 3–2).

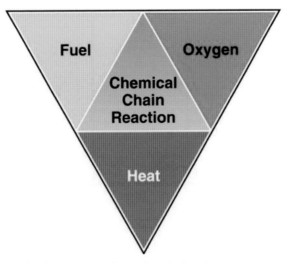

Fig. 3–1. Fire tetrahedron as two-dimensional triangle.

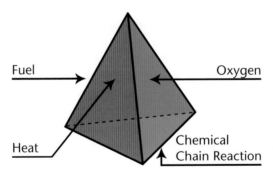

Fig. 3–2. Fire tetrahedron as three-dimensional pyramid.[3]

What they have been trying to tell us all these years, via a graphic that most did not understand, is when you combine the three elements needed for a fire you create a chemical chain reaction and thus fire.

When you look at the fire situation you need to understand the four stages of fire and how fire spreads. According to basic fire behavior science, the four stages are (1) ignition, (2) growth, (3) fully developed, and (4) decay (see fig. 3–3). Understanding each of these stages and trying to determine what stage the fire is in as you size it up helps you to have a better grasp of the fire in front of you.

How fire spreads. Along with the four stages of fire you must know and understand the three ways fire spreads. They are conduction, convection, and radiation, again basic fire behavior science. All three are important and play a role in fire spread,

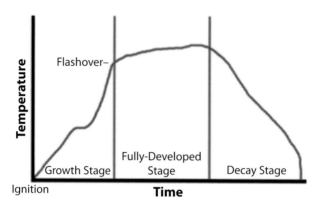

Fig. 3–3. Four Stages of Fire

but so many of our textbooks use archaic pictures to show examples. For example, a camp fire is used to demonstrate radiant heat as it pertains to a fire inside a structure. How many times have you found a camp fire in a structure fire? Nor does a flame to a pipe tell how conduction really spreads fire in a structure. Using examples such as these does very little to show how fire spreads within a building.

The number one way fire spreads is by convection. Yet, be honest, how many really understand convection? Memorizing the definition does not mean you understand it. Holding your hand over a candle does not explain what convection is to firefighters. In the most simplistic of terms, convection in a structure fire is the rising of the fire gases in the smoke. The smoke rises to the ceiling, and then being inhibited, spreads in all directions. The convection currents, while releasing hot gases to the upper levels, draw in cooler, usually oxygen-enriched air at the lower level. It basically becomes a circular current and is referred to as bi-directional flow. The free burning fire creates more smoke (gases) and more heat, giving the gases buoyancy and causing them to rise. The smoke travels through interior open doorways and hallways and rises up to the floors above if there is nothing to stop the action, such as a closed doorway. If the fire continues to burn it continues to release more fire gases and generates more heat.

The gases in the smoke are fuel. If they reach their ignition temperature and have the right air mixture, they ignite. For us as firefighters, this can mean opening up a vent limited fire, allowing the sudden incursion of fresh oxygen-enriched air, causing the fire to flare up and quickly reach flashover! A flashover ignites all the gases within the room, and the flame and heat then follow the gases contained within the smoke to where it might be, and they ignite.

If you are in heavy gas (fuel) enriched smoke on the second floor landing and the room below lights up, you're going to be in a very bad place in a matter of seconds as the gas ignition reaches you. So we must understand and remember that hot convection currents basically follow the smoke. The smoke is full of fire gases and it is convection that spreads these gases. Remember when the gases ignite, they are what will hurt or kill you, so it is critical that you understand convection.

The basics of fire behavior are important when it comes to analyzing risk. Incident commanders and company officers must have this knowledge and continue to keep it fresh. I suggest that every fire department have an in-depth fire behavior review every year in order to keep this knowledge fresh.

3. Have you identified the flow-path?

Over the past decade the term flow-path and the science behind it has been discussed, researched, and overwhelmingly accepted by the fire service. Yet we still have those who don't buy in, which usually indicates they don't really understand the subject, are afraid of change, or are just too stubborn to learn about and accept new or improved thinking. A great example is those who still refuse to wear a hood (yes there are still firefighters like that out there). But the facts, supported by extensive study and research, support the theories being brought forth.

Even though these theories are being presented as new, many are in fact rooted in things of the past. Back in the 1970s I was taught that when it came to cellar stairs you never stood still—you were either moving on the stairs or off them. The reason was that you were in what amounted to a chimney, and the fire could be drawn up the chimney onto you. Think about it in today's modern context of how the smoke travels and exits—that too is a chimney, a way out. Today we call it a flow-path as discussed in Chapter 2. The *flow-path* is the way the gas (fuel) enriched smoke and the heat within the structure are moving and exiting while cool oxygen enriched fresh air enters in bidirectional flow. When we open up a structure for entry, the smoke exits where we opened up. We have created a single flow-path. If we then create a second opening (example: vent a window), we now have multiple flow paths. Convection currents (bidirectional flow) will exit the hot smoke and fuels, and below them, cooler air will flow in. The point where the exiting currents meet the incoming currents is called the *neutral plane* (fig. 3–4).

The fresh, oxygen-enriched air introduced with the venting and opening up helps to fuel the fire and will cause it to grow very quickly and, as we talked about earlier, usually leads to a flashover event. The flow-path is very important for us to know, observe, and understand because when we make entry, if we are in a flow-path, we are in danger of getting caught in a flashover. Underwriters Laboratories have done numerous studies (available online) that show just how fast a fire will grow to flashover stage when we open up.[4] There is no set time, but statistically it is in the two- to three-minute range.

Fig. 3–4. An actual neutral plane at a fire.

4. Is it a vent limited fire?

We now know that with modern building construction, as it relates to the weather tightness and energy efficiency of the building, the fire can be in the growth stage and then become oxygen deprived. As the available oxygen is used up, the fire's growth slows down due to the lack of additional air coming in. This is what we call a *vent limited fire*. Opening the door for entry or venting windows gives the fire a sudden infusion of oxygen-enriched air and causes the fire to quickly take off, grow, and more than likely reach the fully developed stage via a flashover! Having this knowledge can and will help you in your decision-making process.

As part of the new science of fire behavior we are learning about and better understanding vent limited fires. We know that limiting the venting deprives the fire of needed oxygen. It diminishes the flow path and inhibits the fire's growth. When we vent, here comes the oxygen! When we vent we can clearly see the flow-path if it is not already identifiable. There are many ways we can limit the vent and thus help to slow down the fire growth. To start, we need to educate the public that when they flee a burning structure they need to close the doors behind them. Many communities have or are introducing a "Close Your Door" education program within the schools. It's a great place to start. Other such resources are available, such as the Underwriters Laboratories' "Close Your Door" website (closeyourdoor.org). We also need to teach and educate our brothers in blue that if in fact they attempt to make entry before our arrival, it is very important for them to close the door after their attempts, even if the occupants are still trapped in the structure. Recent studies have shown that the civilians have a better survival chance if we limit the vent and slow the fire down or prevent flashover.

Regarding our firefighting activities, we need to break some of our old traditions such as someone *without any orders* running around smashing glass to "vent," thinking they are helping! This is freelancing. Freelancing is a disease that injures and even kills firefighters and trapped civilians. It also causes unnecessary damage to buildings as the fire is able to spread rapidly given the amount of oxygen that is introduced. Many of you are probably saying, "Wait a minute. No one does that anymore." One might think that, but I recently reviewed a fire on YouTube where a member of a mutual aid company came in, and without any orders or directions, took out all the first floor glass. Yes, it is still happening and we need to stop freelancing actions such as this. Do not open the vent until a hose company is in place and ready to go. This means fully geared up, with a charged line that has been bled out and a crew capable of the assignment. The venting can then commence under orders from the incident commander in a controlled and coordinated manner and the door opened for entry. If needed, the company officer can control the door by using a tool or a short piece of rope or webbing to pull the door back into a partially closed but not locked position until ready to go. The goal or objective is to limit the amount of oxygen the fire will get, and hence the fire growth, for as long as possible.

5. What is the smoke indicating?

Many years ago as new firefighters, when it came to smoke, all we knew was that where there is smoke there is fire. We learned that if it was a color other than tan or brown, it most likely contained chemicals. That was the extent of our smoke knowledge then. Today we have learned a lot about the importance of reading smoke.

Some argue that the ability to read smoke is becoming one of the most important things we need to observe at a structure fire. Dave Dodson and his "Art of Reading Smoke" programs[5] are in this author's opinion the industry standard. If you are not aware of Chief Dodson, go online and research him and the subject. A wealth of information about reading smoke is also available on the Fire Engineering website (https://www.fireengineering.com) and in their books and videos.

I cannot emphasize enough that this process, especially because of modern construction and room contents, is becoming a key observation that must be paid attention to. It works well with observing the flow-path. Being able to read smoke assists you with determining answers to key questions:

- What initial actions are needed?
- What number of lines are needed?
- Is it a deep-seated fire?
- What are the heat conditions?
- What is the probable location of the fire?
- Are there flashover probabilities?

Furthermore, understanding smoke helps you with assessing venting effectiveness, knowing if you are containing and controlling the fire, and knowing if the fire is gaining on you.

The most important thing you need to remember about smoke is that *smoke is fuel*. The smoke you see is full of fire gases that ignite. It is the gases that endanger us as firefighters, it is the gases that ignite for a flashover, and it is the gases carried by convection that spread the fire within a structure.

As you learn to read the smoke you become more skilled as a chief officer, a company officer, and a firefighter. Your observations keep you more proactive as you realize what is happening and anticipate what might be going to happen, such as a flashover. To really learn and understand this foundational concept of reading smoke takes many hours of study and practice. Utilize every training opportunity available to you to acquire the skills of reading smoke. Today, firefighters' lives are depending on you to make the correct decisions.

Reading smoke is a key observation when we assess the risks at a fire. When we look at fire and smoke we must pay attention to the four key attributes: volume, velocity, density, and color. Each is different and each is significant in understanding what is happening in front of you. The next section is a primer on the subject, but know that reading smoke is not something you learn in the few paragraphs presented here. It is up to you to learn this subject in depth through study and research (www.fireengineering.com/articles/print/volume-158/issue-9/features/the-art-of-reading-smoke.html).

The four key attributes to observe when reading smoke

Volume. When you look at the smoke exiting, compare the volume of the smoke to the size of the building. The volume gives you an idea of how much fuel is off-gassing. The greater the volume of smoke, the more heating is happening, and the more fuel off-gassing is happening. We know that a smaller structure fills with smoke quicker than a large structure. Ask yourself what is the size of the building you are looking at and compare the volume of smoke. Assessing the volume is the beginning of the process.

Velocity. Velocity is pressure. It is the velocity of the exiting smoke that can indicate how much pressure has built up. Two things cause smoke pressure: volume restriction and heat pushed. Volume restriction causes pressurized smoke to exit the building and quickly slow down. Smoke pushed by heat rises quickly upon exiting the structure. It is heat pushed smoke that we need to pay extra attention to. Along with looking at how fast the smoke is exiting the structure, we need to understand the type of pressure it is displaying: laminar or turbulent (fig. 3–5).

Laminar smoke looks streamlined or smooth when exiting and typically indicates the exiting smoke is not super heated. However, there are times the laminar flow might be a false indicator if the fire is deep seated in a large building and the smoke has traveled a long distance, cooling as it flows to the exit point. We often see laminar smoke as part of the early stage heating, or ignition phase of the fire. We also see laminar smoke as a fire is extinguished and the smoke and steam exiting have cooled.

Turbulent smoke is smoke that is exiting fast and appears to be boiling, angry, or agitated. I often describe it as smoke coming from an old steam locomotive. The turbulence is caused by either high heat or volume restriction. Another aspect of turbulent smoke is that it is enriched with fire gases that are at or near their ignition temperature. We see this in two ways. The first is when turbulent, heavy dark smoke exits; if it finds an ignition source and has the correct air mixture, it will ignite. Second is inside the structure itself. As the fire quickly grows, it will get to the point that the entire room is full of fuel-enriched, dark turbulent smoke. When the different gases reach their ignition point (for example, the auto ignition temperature of carbon monoxide is 1,128°F/610°C), they will quickly ignite and the room or area will flashover.

Turbulent smoke exiting the building can also be used to help locate the fire. Imagine you have turbulent smoke coming from two different windows on the A side, second floor of the structure. The smoke coming from the left window exits, boiling and turbulent, and stays rolling and angry as it lifts. The smoke from the right window also exits boiling and turbulent, but as it lifts it begins to slow down and lose some of its velocity (pressure). Remembering that it is high heat or volume restriction that can cause turbulent smoke, we can look at this example and say that the fire is on the second floor near the left window. The exiting smoke is heat driven and that theory is proven as it stays turbulent, angry, and boiling as it lifts. The smoke exiting from the right window was hot (heat

Figs. 3–5. How is the smoke exiting the building? Is it streamlined or laminar (above), or turbulent (below)?

driven) and perhaps volume driven, but since it was not adjacent to the fire it cools much quicker when exiting, thus slowing down and losing its turbulent velocity

To you as the IC or company officer, reading this smoke tells you that the fire on the second floor is most likely on the left side as you are facing the A side. Bottom line: By comparing the exiting smoke velocity from different openings you will have a pretty good idea of the location of the fire.

Density. Density indicates the quality of burning and the continuity of the fuel. By density we mean the smoke's thickness. Again, knowing that smoke is fuel, we can deduce that thick, dense smoke is gas enriched and if given an ignition source can burn. Think of all the different flashover videos you have watched. What you were seeing were the gases within the dark, dense smoke igniting, filling the compartment with fire. Dense smoke indicates a continuity and abundance of fuel. The thicker the smoke, the more fuel (gas) enriched it is. Years ago, we used to refer to thick, black turbulent smoke as *black fire* and an excellent flashover warning (fig. 3–6).

As an incident commander, think about the crews you commit into the structure. If there is dense, thick smoke, it is almost a sure bet it is spreading throughout the structure via convection. The firefighters who must crawl through that smoke are actually crawling through fuel-enriched smoke.

Color. The color of smoke can tell us a lot. Let's start with white smoke. You see white when water is applied to the fire, creating steam. It is the steam that makes the smoke white. You also see white smoke from early stage heating or the ignition phase. In early stage heating, what we are seeing as white smoke is the release of the moisture in the products as they begin to burn. When most fires start they release some white smoke, depending on the fuel and quality. It might be for a moment or two or it might for a longer period.

Typically, we see smoke in the grey to black range, however. The darker the smoke, the more fuel enriched it is. Remember black fire as described above. The darker the smoke, the hotter it is and that heat is being flame driven.

Smoke from untreated wood gives off brown smoke as it grows through late stage heating, and from there emits black and flame-driven smoke. Stop and think about what and where is the untreated wood in the structure. The untreated wood makes up structural elements, so be observant and cautious. An important caution here: Engineered lumber, which is often used for structural elements in modern construction, also gives off brown smoke as it is heating. The caution here is that engineered lumber such a laminated veneer lumber (LVL) and oriented strand board (OSB) is held together by glues and you do not need direct flame impingement for the glues to heat up and fail. This type of failure can and has caused structural collapse and killed firefighters.

The ability to read and understand smoke is a critical observation point all incident commanders (and frankly all fire officers) need to understand. This brief

Fig. 3–6. Dense smoke indicates the quality of burning and the abundance of fuel.

primer is the tip of the iceberg on this subject. If you are not well versed in reading smoke, you must dig deeper into the subject and make sure you truly understand it. The ability to read and understand smoke guides you in making good and safe fireground decisions and managing the risks firefighters are subjected to in a more informed and safe manner.

6. What is the building construction type?

NIOSH line of duty report contributing factors regarding building construction:

- Deteriorated structural members
- Bow string roof truss construction not recognized by departments

NIOSH key recommendations:

- Train all firefighting personnel on the risks and hazards related to structural collapse.
- Ensure the incident commander receives pertinent information during the size-up (i.e., type of structure, number of occupants in the structure, etc.) from occupants on scene and that information is relayed to crews upon arrival.

We all read and hear about building construction, but do we really understand its impact and danger to us as firefighters? I cannot understand why people are so afraid of learning and understanding building construction as it relates to the fire service. When teaching, I ask students what dangers we have to watch out for regarding building construction and typically only hear about either balloon construction or lightweight truss. Both are good answers, but what about all the other aspects? Firefighters' lives depend on having a working knowledge of at least the basics.

There are five building construction classifications:

- Type I—Fire resistive
- Type II—Noncombustible
- Type III—Ordinary
- Type IV—Heavy timber
- Type V—Wood frame

Let's briefly look at each one.

Type I—Fire resistive. Type I is most easily described as the tall buildings in cities; however, it is a lot more than that. If you live near any urban or suburban area there is a good likelihood you have Type I buildings. We refer to Type I construction as fire resistive. Yes, skyscrapers are Type I, but you might also find it with a five-story building in suburbia (fig. 3–7). With Type I, all the structural components are made of noncombustible materials such as steel and concrete and all components have a two-hour fire rating. All exposed steel is covered with a fire-resistive spray-on material or encased by concrete or gypsum. The exterior walls are not structural in nature but rather attached to the building's structural members, creating a facade of glass, steel, aluminum, or other decorative product.

The number one way fire spreads in a Type I building is by exterior overlapping fire. A recent, tragic example is the 2017 Grenfell Tower fire in London, England.

While most volunteer departments do not have Type I buildings within their community, this type of construction may be within their mutual aid response areas. As an incident commander or company officer, if you can potentially respond to such a building you need to understand its construction and how the fire will spread.

Type II—Noncombustible. The term *noncombustible* refers to the steel structural (bearing) elements of the building only. The actual walls we see are non-bearing and can be of numerous materials. Type II buildings are very common in all our communities and can used for most anything, including single-story warehouses, manufacturing or repair facilities, on farms, as fire stations, and for all types of other functions. This is a very popular type of construction and you must understand it. They are relatively inexpensive to construct and afford the occupants a large, wide-open space. While some are very easy to identify (fig. 3–8), others are not as obvious (fig. 3–9). The second picture shows a Type II building that has been finished to look very different than what we might expect from a Type II.

We need to remember that even though it is classified as noncombustible, it is likely that most everything the building will burn, including furnishings, partitions, the stock or inventory stored within, trucks, heavy equipment, cars, and the list goes on and on. One concern is that steel that is not protected can expand 1 inch every 10 feet at 1,000°F and it will fatigue under its own weight at 1,500°F. An additional concern is what is inside the building. We all know to prepare for the unexpected. For example, if it is a farm building, what is in it that could harm us? Pesticides? Fuel storage? What if it is a small distribution center with pallet racking 20 feet tall? Are there fire protection systems such as sprinklers or standpipes? Know your area and districts to help you manage the risk.

Fig. 3–7. Type I buildings are not just in large cities.

Fig. 3–8. An obvious Type II building.

Fig. 3–9. A Type II building in construction but not in appearance.

Type III—Ordinary construction. Ordinary construction is very common and popular. We see it in buildings over 100 years old and we see it today in new construction (fig. 3–10). It is typically four stories or less but can be found in up to seven-story buildings. Simply put, ordinary construction is when the load-bearing walls are masonry. The older ones are unreinforced masonry (URM) and the newer ones are braced by reinforcing material such as rebar in

Fig. 3–10. A typical Type III building.

concrete or cinderblock. The roof and floors are wood frame. We find this construction used today in retail strip malls and other commercial buildings. From the past, we find Type III ordinary construction in downtown areas— what some refer to as Main Street, USA. With the older buildings, we need to be concerned that the floors and roof are made of wood and when burning can collapse. Years ago, they would do a fire cut (fig. 3–11) on the floor joists and

Fire cut

Fig. 3–11. Fire cut was frequently used to prevent a floor collapse from collapsing the walls.

roof rafters to prevent them pushing the exterior walls outward and causing the building to collapse. Other concerns are the parapet walls, cornices, and other masonry and wood decorative elements. These buildings have collapse potential and we need to proceed with caution.

As incident commander, this is the type of information you need to know and understand as it can make a huge difference in decision making. So many of the older types of these buildings are found in the downtown section of your towns. Usually it is a congested area, with shops, restaurants, sandwich or pizza shops, barbers, and other businesses typically found in our communities. When these buildings burn, knowing how they burn and collapse will help you to keep your firefighters out of collapse zones, and better understand and anticipate how the fire is going to travel.

Type IV—Heavy timber. Heavy timber construction has masonry exterior walls and all the interior structural elements and floors are wood. We commonly identify this type of construction with old mills (fig. 3–12), most of which are from the 1800s. The big difference between Type IV and Type III is that the interior wood components are made of significantly heavier timber. In Type IV you will find large dimension lumber such as at minimum 8×8" posts, 14" beams, and floors that are 2" and 3" thick. Your concerns for this type of building are

Fig. 3–12. Type IV heavy timber construction.

large, wide-open spaces and floors that are oiled soaked from machinery from an era gone by. The old mill type production machines are most likely gone, replaced by newer equipment, or the building may have been renovated for condominiums, apartments, or commercial use. Each of these reuses presents different problems that can affect your firefighters. Know your area and the dangers within.

Type V—Wood frame. Wood frame is our most common type of construction (fig. 3–13). It is used for homes, small businesses, and commercial properties, and today we are seeing a surge in what has been termed *toothpick* construction replacing what may have been Type I or Type III in the past. Within this category there are many subcategories such as balloon construction, platform construction, and post and beam. What each of the subcategories have in common is that they are structures made of wood. Each is different and each has its own dangers to us as firefighters. With toothpick construction, the entire structure is combustible. Add to this the popularity of wood truss and lightweight engineered lumber (fig. 3–14). Lightweight construction adds significant danger and is prone to burn through quicker and lose structural integrity faster. Most engineered lumber is held together by glue. When the materials become heated, the glue melts and loses its intended function of fastening the product together. This means fast loss of structural integrity and faster collapse.

Fig. 3–13. Type V wood frame construction.

Fig. 3–14. Example of some lightweight engineered lumber framing materials.

There are so many variations to Type V that a full description would add pages and pages to this book. A brief warning about lightweight construction (LLW) or toothpick construction (TPC) (fig. 3–15) is needed, however. With LLW/TPC, the entire structure is lightweight wood construction. *Entire four-, five-, and six-story buildings are being constructed of engineered lumber.* Over the past few years there are been numerous fast moving, total destruction fires in these type of buildings, some under construction and some almost complete and about to be turned over to the owners for occupancy. It is becoming more and more a problem and danger to the fire service.

Following are some excellent articles about the dangers of lightweight construction:

- "Toothpick Towers: A Fire Officer's Guide to Operating in Lightweight Wood-Frame Multiple Dwellings" by Glenn Corbett, *Fire Engineering*, 4/1/2018 (https://www.fireengineering.com/articles/print/volume-171/issue-4/features/toothpick-towers-a-fire-officer-s-guide-to-operating-in-lightweight-wood-frame-multiple-dwellings.html).
- "Toothpick Construction: Enough Is Enough" by Jack J. Murphy, *Fire Engineering*, 3/21/2017 (https://www.fireengineering.com/articles/2017/03/toothpick-construction-enough-is-enough.html).

Fig. 3–15. Examples of lightweight (LLW) (above) or toothpick (TPC) construction (below).

Now that the five classifications have been explained we need to go further in our education as it regards building construction for the fireground. I have always suggested the three key observations we need to know are the following:

How is it built? You need to understand what the building is made of and how it is constructed. Is it wood, steel, masonry, or a combination? How will this be affected by fire? Learning about how it is built for all five classifications will give you a good, solid core of understanding the building. You need to learn the strengths and weaknesses of each classification as it affects us as firefighters. Things like cocklofts, parapet walls, steel directly exposed to heat and fire, and collapse potential. All of this is key information to help you to be proactive and safe.

How does it burn? With each classification the fire will behave differently. How will the fire spread and what affects its path of travel? You need to understand how the type of construction will inhibit or help to spread the fire's travel. The classic example is of course balloon construction. Each classification needs to be understood. A Type III ordinary building will have cocklofts and possibly shared pockets in the masonry walls, allowing fire spread. Fire will behave differently in a Type V wood frame platform constructed building than it will in a Type IV heavy timber. You must understand these basics.

How does it fall? How often do we hear people talking about the collapse zone and then at a fire, totally ignore the cautions? The collapse potential is not just the walls and parapets falling on you. It also includes many other elements such as a floor of lightweight construction burning and collapsing under your weight or the roof rafters that are trusses collapsing while you are trying to cut a vent hole. If you need verification, review some of the NIOSH LODD reports.

Depending on where you are located and what your mutual aid response area is, many of us can face all five classifications. For other volunteer organizations, we might only see Types II, III, IV, and V. But regardless of where you are, you must know the basics of building construction for your response area.

As a competent officer and incident commander, to manage or reduce risk to the firefighters you need to understand how the building before you is built, how it burns, and how it might collapse.

There are numerous books and classes available on building construction. Take the time to expand your knowledge and be prepared to make decisions that affect risk to your firefighters based upon facts.

Learning about this subject is not done just by reading an article or watching something online. You need to spend many hours reading, studying, taking classes and seminars, and more. Failing to do so is a potential disaster in the making. Lack of knowledge on your part during an incident can increase risk to those operating on your fireground. Be proactive and pay attention to this subject so we can be safer. I'd like to recommend the book *The Art of Reading Buildings* by Mittendorf and Dodson[6] as an excellent resource to begin the study.

7. Are civilians trapped or missing?

Scenario: Upon arrival, you are told of civilians trapped or unaccounted for. This presents a danger and risk to your firefighters. First, you need to determine whether the occupants are in immediate danger. If so, you need to know their location within the structure and their likely condition. These are tough issues to talk about. Many volunteers are from smaller communities where we know our neighbors. In some communities, it is almost a sure bet that one of the responding firefighters will know the home's occupants. We cherish small-town living, but in an instant such as this it can become your worst nightmare. With any rescue or search, we need to identify fire and smoke conditions, the possibility of structural collapse due to lightweight construction or prolonged burn time, and any unusual conditions that could affect the rescue attempt. (In Chapter 4 of this book we review the survivability profile.)

In a rescue situation, one of the biggest risk assignments that you can make is to what company the tactic is assigned. You have to know that the officer and crew are capable, so this is a critical decision. Remember, we are not all the same; skills and abilities vary. The assignment of search and rescue for a trapped occupant is dangerous. It is a skills and guts demanding job. Choose wisely! If the assignment is made via radio to a company en route, but upon arrival, you see that it is not the right crew for the assignment, the only choice is to transfer the risk to a crew that is capable. Your decision might be based upon their physical abilities, skill levels, or training. Whatever it might be, if they are the wrong choice, replace them.

8. What is our ability to communicate on the fireground?

NIOSH line of duty report contributing factors regarding communications:

- Ineffective fireground communication as it regards the incident action plan
- Lack of common radio frequency

NIOSH key recommendations:

- Ensure that firefighters communicate interior conditions and progress reports to the incident commander.

Examine the communications situation. There are many factors involved. For example, is everyone on one frequency and able to talk to each other? On the fireground we need all the firefighters, especially the company officers and incident commander, to be able to communicate via radio. Does every member of a company operating in, on, or under the building have a radio? The radio is an important management and safety tool. It allows the command staff to communicate with the company officers and allows the company officers to provide feedback and benchmarks to the IC. All company members should have a radio, primarily as a safety tool. If they become lost, separated, injured, or trapped they can immediately call a Mayday.

What about when mutual aid companies are operating on your fireground? Can we all communicate via radio? If not, how are orders transmitted, feedback provided, and the safety of all maintained? In 2003 a volunteer firefighter died in a basement fire. There were numerous fire departments, mostly adjacent neighbors, but they were almost all operating on different radio bands and frequencies. Bottom line, they could not all listen and/or talk to each other. This became an issue when it was determined that a firefighter was missing and most likely trapped in the basement. There was no RIT on scene. In spite of gallant efforts by some firefighters who took the initiative to find and rescue him, the firefighter died. The NIOSH report cited communications as a factor.

The following is an excerpt from the NIOSH LODD report:

> *NIOSH Report F2004-02*
> *Date Released: October 26, 2004*
> *Basement Fire Claims the Life of Volunteer Fire*
> *Fighter—Massachusetts*

In the report it states numerous recommendations including the following as it relates to communications:

> *Recommendation #7: Municipalities should ensure that companies responding to mutual aid incidents are equipped with mobile and portable communications equipment capable of handling the volume of radio traffic and allow communications between all responding companies within their jurisdiction.*
>
> *Units responding to or engaged at incidents should have the necessary radio frequencies/channels to be in contact with other units providing mutual aid. These units should also have the capability to monitor the fireground activities while en route. During this incident, many of the units could not communicate with the IC or the local dispatch center on either their portable or mobile radios.*

To reduce risk on the fireground we must find solutions to guarantee that we can all talk to each other when operating together.

Another issue, and probably the number one reason given for not issuing radios to all members, is the constant out-of-control radio traffic. Do many, if not most of us, have problems with this? The answer is yes. But, let's look at it from a different angle. Why is radio traffic out of control? The answer seems to be because we have enabled and allowed it. How many departments run drills on how to use a fire service radio? I can hear many of you laughing as you read this. And yes, that is the reaction I expect. "What do you mean radio drills? All you have to do is show them how to turn it on, select the correct channel, and then 'push to talk.'" If this is your thinking, and I guarantee it is a very common response, then I say to you "No wonder your fireground radio traffic is out of control." We first need to teach our members the importance of the tool (radio), its importance in safety (Mayday call), and its importance to the command staff and company officers for communicating tactical assignments, feedback, and progress reports for benchmarks. From there, we need to show them (yes you actually have to talk and role-play) how to transmit a brief but descriptive message, what messages they should be sending, and what they *should not* be sending. Unless they are calling a Mayday, there are very few reasons that company members need to clog the airways. Legitimate and useful transmissions include when the officer splits the company and you have need to share or ask information. It might be when, as a hose company, you are spread out over two floors, stretching the hose to the seat of the fire. There are many other legitimate reasons. We need to teach and show them how to use the radio, including what type of transmissions to make and how to verbalize them. Specific and brief usually works best. If problems of excess radio traffic continue, you need to call out the offenders and reprimand them.

The ability for all firefighters to communicate on the fireground via radios is critical. But out-of-control radio traffic creates an extremely dangerous risky environment, endangering the lives of those operating on the fireground.

9. Is the incident command system (ICS) being used?

NIOSH line of duty report contributing factors with regard to ICS:

- Ineffective incident command (IC)
- Lack of an incident management system (IMS)
- Ineffective incident management
- Incident management system not implemented at the fire scene
- Ineffective span of control
- Initial arriving units not establishing/performing/implementing an incident management system, an overall incident commander, an incident action plan (IAP), and a 360-degree situational size-up

NIOSH key recommendations regarding ICS:

- Fire departments should develop, implement, and enforce a written incident management system to be followed at all emergency incident operations.
- Ensure the first due company officer establishes command, maintains the role of director of fireground operations, does not become involved in firefighting operations, and ensures that incident command is effectively transferred.
- Ensure the first due arriving officer maintains the role of incident commander or transfers command to the next arriving officer.
- Develop, implement, and enforce a written incident management system to be followed at all emergency incident operations and ensure that officers and firefighters are trained on how to implement the incident management system.

In the early 1990s, the incident command system (ICS) became the fire service standard for command and control at fire scenes. It was not mandated but many of us saw the importance of using a system that clearly defined structure and hierarchy at an emergency scene. In 2004, the U.S. Department of Homeland Security developed the National Incident Management System (NIMS), which they integrated into the ICS. The use of the integrated NIMS and ICS was then mandated by the federal government for all emergencies (Homeland Security

Presidential Directive 5, issued Feb. 28, 2003).[7] Failure to do so would exempt the community from federal funding. This gave us a single, standard approach to incident management. It clearly defined positions, who was in charge, and gave us common terminology. Bottom line, no matter where we went, all the fire departments and emergency service personnel would be in unison with regard to command structure and terminology.

Today, we are still struggling with getting all of us on board with this concept. It is hard to imagine or believe, but there are still organizations that couldn't be bothered. Perhaps it is change they are afraid of, perhaps they think that conforming is taking away their power and authority, or perhaps they are just ignorant as to the importance of this system. Whatever the reason or excuse given, it's time to get on board and use the ICS system at every incident.

ICS is not just pulling up on the scene and establishing command. It is learning to work within the structure of the incident management system, learning to keep within a span of control as it regards direct reports, and learning to use the system to assist you as the incident grows in scope. Do you use mutual aid? Are you called out for mutual aid? Then common terminology is very important and critical to our safety.

The incident management system will make you a more effective IC, give you better control of the overall incident, and keep your members safer. Use it.

10. Is there a RIT in place?

NIOSH line of duty report contributing factors regarding RIT:

- Rapid intervention crew (RIC) procedures not followed and/or implemented
- Lack of rapid intervention crews
- Lack of assigning rapid intervention crew (RIC) or firefighter assist and search team (FAST)

NIOSH key recommendations:

- Ensure that a rapid intervention team or crew is established and available to immediately respond to emergency rescue incidents.
- Ensure that a rapid intervention team (RIT) is established and available at structure fires.
- Fire departments should ensure that the incident commander establishes a dedicated rapid intervention crew (RIC) or crews and that the RIC is available throughout the incident.

First off, let's pause and clear the air on terminology. A *rapid intervention crew* (RIC) is considered in the world of ICS the correct acronym as a crew is defined as a company that is not geographic in nature or assignment. Yet we use FAST (firefighter assist and search team), and RIT (rapid intervention team). Where you live seems to be what influences the words we use. Frankly I could care less what you call it. In this text I use RIT. What I care most about is that you have a trained and capable RIT on-scene and ready to go. If your focus is more on correct terminology rather than making sure your members are truly RIT capable, then you are endangering your firefighters. *It is their skills that matter.* I don't care if it is a crew or a team, but I do care that they have the skills to attempt to save my life.

A rapid intervention team is not traditionally considered a part of risk management. In fact, you will not find it mentioned in any of the management guidelines that I researched. Technically, this is correct as managing risk is identifying the dangers and trying to limit or control them. But then we have the issue that you have done your job well and made great efforts to limit or control the risks the members are exposed to, but something goes very wrong and a firefighter is down, whether it be lost, trapped, injured, or unaccounted for. We have a firefighter in trouble with a very limited air supply, in a very toxic environment. The risk and probability of injury or death to the firefighter is very high. If there is not a RIT, or worse yet the RIT is in name only, staffed to fulfill the standard by untrained, incapable people who will not be able to fulfill the mission, then what? RIT is all about training and dedication to saving one of our own.

> **Having a capable and well-trained RIT is part of managing risk.**

11. Is the weather a factor?

> NIOSH cited key recommendation regarding weather:
>
> • Be prepared to use alternative water supplies during cold temperatures in areas where hydrants are prone to freezing

As you assess your dangers and risks, consider the weather. It can and does have a profound effect and influence on firefighter risk and safety. Let's look at a few extremes. Very warm weather affects our performance and endurance. Extreme heat and humidity, in parts of the country where it is not the norm, has additional adverse effects on operating members as they are not used to it. We must anticipate this. Even as volunteers, it is very important that we preach and push hydration as an everyday action that must be taken in order to be prepared. Firefighters in our southern states are more used to heat and humidity, but they also are exposed to the adverse effects. When you respond to a fire in weather like this, get more help early. Don't delay. We need our firefighters to be able to exit the structure and rehydrate for their safety and health.

In the parts of the country where extreme winter weather happens, we must deal with freezing temperatures, snow, and ice. This type of weather affects our response times and fireground operations. Like extreme heat and humidity, sub-freezing temperatures have a detrimental effect. It causes hydrants and lines to freeze up; creates icy, slippery roads; and creates icy conditions on the fireground that can cause firefighters to slip and fall. All of these are fireground and operation dangers and risks. If you are in an area that runs tankers (the ones with tires that stay on the ground, not fly) extreme cold creates very slippery conditions at the fill and dump sites. If your tankers slosh and spill water as they respond, it creates icy road conditions that are hazards to all. At the dump sites, ice creates conditions where extreme caution needs to be observed. If the weather is snowy or there is heavy rain, then our tanker response and water supply shuttle times increase significantly as they must slow down and adjust to the poor road conditions. Get more tankers and make up for the long shuttle times by increasing the volume of trucks and water on the road.

Anticipate what effects weather can have on your incident and the members operating and start more help and equipment early.

12. Is water supply an issue?

NIOSH line of duty report contributing factors regarding water supply:

- Inadequate water supply

NIOSH key recommendations:

- Fire departments should ensure that an adequate water supply is established and maintained.

To begin with, a lack of water puts the firefighters' operation at risk, especially if you are choosing an interior attack. Many volunteer fire departments are blessed in that they are in an area where there is a good and plentiful municipal water supply. They are usually found in more suburban areas. And then there are the rest of us. So much of rural America is protected by volunteer fire departments where water supply is an issue. What if you have a large industrial structure that is burning out of control? A personal example is a fire we fought in 2007—the Burnet Mill fire in Uxbridge, MA (fig. 3–16).

For the first three hours, the fire attack used the municipal water supply, but we were quickly depleting the town's water supply tank and had to get off the hydrant system before we collapsed the tank. At this point, a large part of the community had no water to their homes. We went to draft in two of the three rivers that flow through the town. At its height we were flowing 20,000+ gpm at draft. If you are in a community that has these types of exposures and buildings, water supply could become an issue for you.

For all the departments that live and die by the tanker shuttle, you must, as part of your pre-plans and run cards, have a plan in place via mutual aid to get the water flow that you are going to need. The two biggest issues with a water shuttle are the initial time needed for the mutual aid response and having enough tankers in the shuttle to deliver the water flow required.

Be smart, be proactive, and plan ahead.

The answers from these twelve points will help you to identify the risks and dangers we are exposed to. The answers will help you to prioritize what needs to be done and what staffing and capabilities you have.

Fig. 3–16. The Burnet Mill fire (Uxbridge, MA) quickly depleted municipal water resources forcing an entire drafting operation. (Photo courtesy of Ken LaBelle)

Prioritizing the Risks and Dangers

As the incident commander you must now take the information you have gathered and logically prioritize them for action. Our priorities to ourselves, our firefighters, and our community are always:

1. Life safety
2. Incident stabilization
3. Property conservation

Prioritize every decision you make in this order. These three priorities are and have been for quite some time nationally accepted and used by the fire service. It has always puzzled me when teaching an officer's class how many still don't know or understand this information. Get on board. It will help you to reduce risk by applying some knowledge and common sense.

In an article published in *Fire Engineering* in March of 2015, author Richard Ray[8] focused on the volunteer service and wrote:

> *Initial actions performed by first-arriving members are the corner stone of accomplishing fireground priorities such as life safety, incident stabilization, and property conservation. However, without the appropriate staffing and resources, critical tasks/actions will not be performed, which leads to failure on the fireground in the form of injury or death and loss of property.*

Too often, this is our reality. We know what needs to be done but we do not have enough resources to get it done in a timely manner. If this is your situation, then you must abide by a risk management system and use these three incident priorities to help you deal with the situation as best and as safely as you can.

As you evaluate the three incident priorities, you must determine the severity or likelihood of the dangers and risks happening. You must also take into consideration your current staffing level, their capabilities and training, and if they really are the right firefighters to deal with the situation before you.

Life safety

Your size-up and assessment of the dangers and risks may determine there are trapped or missing civilians. Life safety is our first priority, but the incident commander and all officers must remember that our lives—the lives of the firefighters—must come first. For many years, it was expected that firefighters would lay down their lives, regardless of the risks, no matter what. I remember being told by a firefighter from a big city that it was our job to die if needed. That may have been the pledge then, but this is now. Today, our job is to remember that we all go home. This applies to fire departments, firefighters, and emergency responders. Yes, I will take risks as needed, but they will be calculated risks that, based upon my training and abilities, will give me a high probability of survival. Giving your life to save a victim long dead is not calculated risk. Each situation will be different. Later in this book we talk about the survivability profile and how it helps in analyzing the inherent risk of a rescue.

If it is determined that we have the chance to affect a rescue, then we must look at and evaluate the available crews' abilities and training. Are they capable? Using a crew that is not truly capable or is lacking in training is in effect increasing the danger to them as it regards life safety.

An important part of prioritizing the risks and dangers of life safety is evaluating the fire situation. You must understand fire behavior, the science of modern fire behavior, and know how to read and understand the smoke in order to evaluate the life safety situation.

What is the severity of what you are looking at right now? Life safety for us versus civilians is different. For us, it is evaluating the severity of the situation, the inherent risks we will be exposed to, and the level of training you and the companies operating have. Bad decisions lead to bad things happening! Trained and capable firefighters make a difference in your evaluation. Remember we are not all the same. Our capabilities and training vary. This is a key point in evaluating the risks and dangers to the firefighters. For the civilians, we must look at what their current situation is. Are they lost, trapped, unaccounted for? Do we know where they are within the structure based upon their location, adjacency to the fire and toxic smoke, and how long they have been exposed to it? You must take these into account in order to evaluate the dangers they are exposed to and if it is a prudent move to attempt a rescue. The answer is not always going to be yes.

Incident stabilization

Incident stabilization is what needs to happen to minimize the damage to the structure. At a structure fire, simply put, our goal is to put the fire out and minimize the amount of damage being caused. We stabilize the incident by taking the correct actions needed. This includes but is not limited to choosing the right tactics, initiating fire suppression, ventilating, and assigning only companies that are capable of achieving the assigned tactic. As volunteers, a significant challenge we face with doing this is the immediate resources on hand and how long it will be before we will have more in order to address proper incident stabilization. Remember that we need to address the problems identified, and then assign resources in a logical, safe, coordinated manner. You cannot assign what you don't have, or to those not capable of doing the task safely and properly.

Property conservation

When I say "property conservation," what pops into your mind? For most of us, it is probably salvage and overhaul. This is partially correct, but property conservation is a whole lot more. It is the care we take when moving a hose line through the first floor of a house, not on fire, heading to a bedroom fire on floor two. It is using the correct tactics to get to and knock down the fire quickly. It is coordinated ventilation, using the right stream, and more. As volunteers, property conservation is also a great public relations tool. It is committing members, once adequate help is available, to gather furniture in a room below the fire and covering with salvage covers to minimize water damage from above. It is minimizing unnecessary damage to windows by controlling and preventing freelance venting. Homeowners always remember that the firefighters went out of their way to help protect the belongings and furnishings within the structure. They also remember things getting trashed that did not need to happen. These people pay taxes and they vote. Good public relations benefits both us as firefighters and the public.

In this chapter we covered the importance of identifying the dangers and risks of the incident. It is important that we identify these dangers/risks and prioritize them. Our members and all those operating at the emergency scene are placing their lives in our hands. Do your job so that everyone can go home.

Control, Reduce, and Eliminate Dangers

4

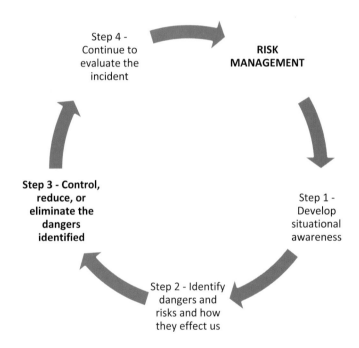

Step 4 - Continue to evaluate the incident

RISK MANAGEMENT

Step 1 - Develop situational awareness

Step 2 - Identify dangers and risks and how they effect us

Step 3 - Control, reduce, or eliminate the dangers identified

In Chapter 3 we examined how to identify the dangers and risks we are exposed to. We looked at the importance of prioritizing them and how our own life safety is paramount. In this chapter we examine how to take the information gathered and turn it into actions to help to control or eliminate the dangers we are exposed to. It answers a lot of questions about how our organization operates. Along with the questions, I challenge you with specific thoughts and ideas that should give you a better understanding of why I ask the questions and why the answers are very important.

Step 3. How Do You Control, Reduce, or Eliminate the Dangers Identified and How Can You Reduce the Risks to the Firefighters?

To begin step 3, I always suggest that first we go back to the very beginning, before the next fire or call, and start with pre-incident questions. We examine something different, something that most do not think of as pre-incident: How has your organization performed at past fires and incidents? When teaching I like to ask my students, "If I observed one of your fire scenes, what would I see?" A proactive leader must ask honest questions about how that organization operates and performs. How do you *really* operate at a fire or incident? You must accept only honest answers, because fooling yourself is a very foolish and dangerous game to play! The answers give you a true assessment of your organization's abilities, strengths, and weaknesses, along with what needs to be fixed or improved. The answers help you to realize what is done in accordance with safe practice and what needs to change. Our good and bad habits affect what happens on the fireground. The bad habits and practices work against you always, while the accepted or safe practices help you with your risk management. As we go through this next segment I provide some direction and questions you need to ask to better understand how you operate. The answers will assist you as you begin controlling the dangers.

Let's look at three key operational observations:

1. Do we have the proper equipment?
2. Do we have the proper knowledge?
3. Are our personnel adequately protected?

Start, as suggested, with past incidents. After you have gained that honest knowledge you have to decide is there room for improvement or are you doing everything perfect. Choose, but choose wisely. With your past incident knowledge, review the fire or incident before you right now and ask the same three questions as they pertain to what is happening right now.

Observation 1.
Do we have the proper equipment?

Equipment is a lot more than the apparatus. We need to look at operational habits in a much deeper way. To assist in the analysis of "Do we have the proper equipment?" let's look at how we have conducted ourselves at past incidents. To do so I suggest eight areas to review:

- Proper personal protective equipment (PPE), including self-contained breathing apparatus (SCBA)
- Proper hose/nozzle configurations
- Fireground communications and radio use
- Use of a thermal imaging camera (TIC) when making entry
- Laddering of the building for safety and firefighter egress
- Use of gas meters
- Proper apparatus
- Water supply

Each revolves around equipment, but is much more than the physical equipment. Let's look at these eight areas individually.

Proper PPE, including SCBA. In the twenty-first century this should be a no brainer: Are all our members on scene in full PPE? Is the PPE current, clean, and inspected frequently? Is it being worn properly? Everything buckled, snapped, Velcro joined and sealed, collar up, hood properly worn, helmet chin strap where it belongs? It's a chin strap, not a brim strap. Wear it properly and your helmet will always stay where it belongs. What about SCBAs? Are all members who could be committed to the incident, other than those in a support role, wearing SCBA? Is it worn properly, waist strap secured snugly, shoulder straps slightly loose? Are the members coming off the apparatus with full cylinders?

Does your department have a standard operating guideline (SOG) that states SCBA will always be worn and operating when entering any IDLH (immediate danger life and health) environment and if so, do all your members abide by the SOG? Or do you have those who think the rule doesn't apply to them? It should be no surprise there are still firefighters out there who don't follow common sense when it comes to SCBA. Doing so puts them at risk, and not wearing their gear properly can create a danger to the members of their company if things go bad.

Is your turnout gear in compliance with NFPA 1851?[1] Are you aware that our gear now has a shelf life of 10 years and then it must be removed from service and replaced?

> *NFPA 1851 §10.1.2. Structural firefighting ensembles and ensemble elements shall be retired in accordance with 10.2.1 or 10.2.2, no more than 10 years from the date the ensembles or ensemble elements were manufactured.*

With these changes come hardships for many of us. It is very important to clean and maintain turnout gear so it can function properly and protect us. To do this, your members need to wear their gear properly, when it needs to be worn.

What about your SCBAs? Are they cleaned and maintained? Are the bottles hydrostatically tested as required, and are they within the date of the cylinder life cycle? We know it is important to clean our PPE in the extractor to remove particulates and toxins, but do you clean the SCBA straps and other elements? You should. If I climbed into your cabs right now, would I find all the SCBA cylinders full?

Maintaining your PPE is a good first step in asking if we have the proper equipment.

Do we use the proper hose/nozzle configurations? Selecting the correct line is a critical first step in any fire attack. Yet how many departments on arrival, regardless of the fire and what they know or don't know about the fire and the given situation, still pull a preconnected 1¾" line? Why does this happen? I think it is back to the standard answers, "That's what we always do" or "That's the way we do it here." Not good answers, and it shows a lack of understanding of what dangers different fires expose us to. Every fire is different. House fire, mill fire, car fire? The 1¾" line is not the answer every time. We are talking about trying to control or eliminate the dangers we are exposed to and that means looking at the fire situation and determining the correct line to pull! When you look at and consider what size hose to advance, you need to understand possible fire flows and how to estimate them. The National Fire Academy (NFA) has a simple formula for residential structure fires: length × width divided by 3 = needed gallons per minute at 100% involvement.[2] For example, you have a 50 × 30 single-story house. So, 50 × 30 = 1,500 divided by 3 = 500 gpm at 100%. If 25% of the structure is involved (a bedroom for example), you would need 125 gpm (25% of the 500). If it is a two-story home, then take the 500 gpm and double it to 1,000 gpm. This is an easy-to-use formula that will get you into the ball park. Besides the NFA formula there is the Iowa formula. The NFPA has their own formula and it is contained in the NFPA 1142, *Standard on Water Supplies for Suburban and Rural Fire Fighting*, 2017. The formula is very involved and is specific to building types, sizes, contents, etc. but it will help you to determine what exactly you should have for a water delivery rate.

NFPA 1410, *Standard on Training for Initial Emergency Scene Operations* is a great tool to develop training drills that will measure your department's performance.

Chapter 8 Required Performance for Handlines states that we need to flow an attack and backup line with a combined flow of 300 gpm minimum.

8.2 Required Flow

8.2.1 The total flow of the required streams shall be a minimum of 300 gpm (1135 L/min).

8.2.2 The initial attack line shall provide a minimum flow of 100 gpm (400 L/min) from the nozzle.

8.2.3 The required flow from the backup line shall be a minimum of 200 gpm (750 L/min).

Some of the training/evaluation evolution provided in 1410 Annex A show suggested layouts of two lines advanced and flowing in a limited amount of time. For many of us, stretching two lines within 3 minutes of arrival will be difficult or impossible due to staffing issues. However, we should strive to stretch and flow in a very short amount of time. **Fast water on the fire is the goal! Again, these are excellent performance based drills, that will assist your department and its training.**

Obviously other types of buildings containing more flammable products such as oils and solvents will compound the issue, as would a fireworks or gunpowder factory. Know your area, know what is needed for water delivery rates, and then know what you can actually achieve.

Another very important component of the hose configuration is selecting the correct nozzle. The hose and nozzle configuration you select is critical in supplying the gpms expected. In my years of visiting many departments and teaching I have seen some hose/nozzle configurations that showed a lack of knowledge on the department's part. Here are two real examples:

Example 1. The department had decided that all this talk about using smooth bore nozzles was a good idea. So, they took their 2½" combination play pipe nozzle, removed the combination tip, and then went into the back room where they stored all sorts of old tips from bygone days! They found a ¾" tip (118 gpm @ 50 psi), which they put on the 2½" play pipe. Friends, we all should know that when you pull a 2½" hose you are expecting a flow of 250 to 360 gpm. But, when it comes to smooth bore pipes, we need to remember that size matters. A 1⅛" tip is designed to deliver 265 gpm, while a ¾" orifice (the one they had lying around and decided to use) will deliver around 118 gpm. What this means is that you are stretching and pulling a 2½" line thinking you have a good large flow when in reality you are not even flowing half of what you think! We need to know and understand tip sizes, gpms, and nozzle tip pressures. Obviously they did not possess the proper knowledge to know this.

Example 2. This department had some of their 1¾" attack line fitted with old 1½" constant flow nozzles. They did not understand or realize that the numbers stamped on the stem indicated flow. In this case the number was 60, so they were initiating a fire attack with a 1¾" hose line, capable of delivering 150 to 180 gpm (with the correct nozzle!), but in this case the 60 indicated the nozzle flow was 60 gpm. This is not in any way a safe flow for fire attack.

Gallons per minute matter. Examine your hose and nozzle configurations and make sure they are a proper match and do the job you are expecting. Most nozzle manufacturers can supply you with a chart that shows you the actual nozzle flow when operated at a different nozzle pressures. These are a good resource. Determine what you might need for water on hand and plan for it either with hydrants, tankers, or a combination of both.

How is your fireground communications and radio use? What is your fireground communications like? If we listened in on one of your calls, is the radio traffic calm and controlled or is everyone yelling and trying to transmit at the same time, causing nothing to get through? Is pertinent information transmitted in a timely manner, or is there so much radio traffic that officers cannot get a clear opening to send an important message? Do all members operating within the structure have radios? If not, why?

When I ask firefighters to describe their department's fireground radio traffic, the most common answer (or complaint) I hear is that it is out of control. I agree with that assessment. But what we need to understand is *out-of-control radio traffic is our own fault.* I'm sure that statement will cause some of you to pause and wonder what I am talking about. Let me take one more step and tell you that it is my deep belief that all members of your department should be issued their own portable radio. If that is not possible, then all members whose company has an assignment that puts them on, in, or near the fire building must have a radio. Now that statement I know has caused many of you to think I am fool and don't know what I am saying! Well, you're wrong. I'm thinking about safety and operational management with that statement.

To help you understand where I am coming from, let's go back to the beginning. What has your communications training been like? Probably like most it goes like this: "This button is on/off. Here is how you change channels. To operate, press this button (key the mike), pause, and then talk. When you're done, let the button go." You call this communications training? What this teaches is how the radio works, not how to communicate, when to communicate, and what to communicate.

I think of communications training differently. Yes, we need to teach how the radio works and its operational features. However, next we need to teach the purpose of the radio and how that purpose changes with different functions and positions within the ICS. For our new members, they need to see the radio as a personal safety tool. If they get in trouble they can call a Mayday. Mayday training is a high priority. Our members need to understand when and how to call a Mayday, an international distress signal meaning to "come and help me." For us in the fire service, I take the liberty to redefine it as "help me or I'm going to die." Members need to be taught that officers use the radio to receive and give orders, report conditions, make progress reports, request additional help, or send urgent messages as needed. An example of an urgent message would be to report an interior stairway has collapsed or burnt through. Teaching all our members *who* should be talking, *when* they should be talking, and *why* they should be talking on the radio will reduce radio traffic. If certain members abuse the radio in direct contradiction to their training, they should be spoken to and, if it continues, disciplined.

The bottom line here is that most times out-of-control radio traffic is caused by allowing it to happen and by not properly training our members on the purpose and importance of this tool and why misuse is dangerous and unwanted. If you think the solution to great communication abilities on the fireground is to limit radios to only certain members, you're wrong. And, if you implement that practice and a firefighter gets lost or trapped and cannot call a Mayday quickly, you might be dead wrong!

As with most skills in the fire service, training is the key to excellence. Train your members properly on the use and purpose of radios and you will get better results.

Do your company officers use a thermal imaging camera (TIC) when they make entry? When TICs were first introduced to the fire service, they were very expensive and many fire departments were not able to purchase one. Over time the price dropped and today almost all departments have them. Many departments have a TIC on every piece of apparatus. Today, some manufacturers offer a TIC built into the SCBA, making it become a standard piece of equipment.

How do your department officers utilize the TIC? I still see many officers not bringing the TIC in with them. They complain that it is too cumbersome, or they have enough to carry, or they just forget it. Well, note to self, the TIC is part of a good line officer's toolbox. The TIC is a very important tool for the officers making entry. It will help them to better understand what is happening inside, look for missing or trapped civilians, locate the seat of the fire, check on their crew's personnel accountability report (PAR), see fire extension, and a host of other things.

Part of the problem here is the training that we have conducted with this great and important piece of technology. What has your department's TIC training been like? Do we do the usual and show firefighters how to turn it on and then scan and look at each other on the apparatus floor? Sound familiar? Please remember that the TIC, in black and white, shows hot as white and cold as dark or black. On the apparatus floor, this means the firefighters will appear as hot, representing the fire. This is directly opposite of what you will be looking for in a fire! We are training our brains to recognize something that is incorrect! Do you train by putting your hand on a metal surface and then showing everyone your handprint? What has that accomplished? We don't look for handprints in a burning building! Some people use the kitchen stove by turning on the burners and showing everyone the "fire." How is this type of training applicable and achieving the desired visual and mental memory needed? Are the members trained on how to change the battery? Is it taught in broad daylight, without structural gloves on? That's not the realistic conditions that will have to be overcome. Train to achieve this skill, blacked out and with structural gloves on. Only then will you achieve the desired skill set.

Using a thermal imaging camera as a company officer is not progressive. Fifteen or twenty years ago it was progressive, but today it is the norm and accepted best practice. We need to make sure our officers and all members are properly trained in how to operate a TIC and, more importantly, how to interpret what they are seeing. If you have access to a burn building, utilize it for TIC training. As of this writing I am aware of one TIC manufacturer (MSA) that offers a training mode setting that reverses the image colors. This allows you to practice if you are not in a burn building and see a more true representation of what you will be seeing and trying to interpret in a burning structure.

The use of the TIC by company officers operating within the interior is an important skill that provides a higher level of safety, reducing risk to some degree. We need to use it always and to continually train on its use.

Do you ladder the building for safety and firefighter egress? When you have firefighters operating on the second floor and above, do you ladder the building to provide them with emergency egress if needed? Almost every time I ask that question, I get a variation of the same answer: "Ladder the building? We don't have enough people to put the fire out." Yes, a fair answer, but then I ask: As soon as you have enough help, do you throw ladders for firefighter safety and egress? The typical answer is still no. Why? We never needed it before is a weak excuse. It reeks of complacency. Complacency is showing that you are in a happy place with what you do, clinging to only what you know or think, while ignoring the dangers that surround you and the firefighters. This attitude and lack of action has injured and killed firefighters.

Do you have an aerial? Even if it responds with a driver/operator only, upon arrival, is it positioned and set up, stick out of the bed, ready to go? A single firefighter with some skill should be able to achieve this quickly. Now, if we need the aerial, precious minutes are not wasted doing what should have already been done. If the operator cannot achieve this, then we need to consider is the person incapable or is more training needed. An aerial without a capable and trained operator is useless, and this uselessness makes the aerial useless. Training, good anticipation, and planning helps to get good results.

Laddering the building is not an outrageous task at a typical dwelling fire. A single firefighter can raise a 24' ladder to a second-floor window. When firefighters enter a burning building, you need to evaluate the dangers they are being exposed to and ask yourself (think), if they get in trouble, how can they get out? A ladder or ladders is a way out. When someone is screaming Mayday, it is not the time to look around for who can raise that ladder. As for enough manpower, I have another solution for you to consider. Why not be more aggressive with your mutual aid and call for more help right away? We all know two things: For most of us, initial staffing is an issue, and our neighbors have the same problem. So why not set up automatic response with your neighbors where more help from all is initiated with the initial 911 call? This increases your initial staffing and theirs right away.

The excuse of not enough people seems to be used for everything. Not throwing ground ladders or positioning the aerial for egress and escape is not having the proper equipment ready for when it might be needed.

How do you use your gas meters? We all have gas meters now. Most of them were initially purchased for carbon monoxide (CO) calls. Many of us went on to add to the capabilities of our meters for sulfur dioxide, cyanide, and other poisonous gases. Do you use these meters to determine when it is safe for our firefighters to take their SCBA off? If not, you should! Many years ago when I started it was the custom to take our masks off for overhaul once the fire was knocked down. We had no clue that even then, the lingering smoke had poisons in it that were dangerous. Times have changed and firefighters today are taught to never go off air unless the toxic environment has been deemed safe. This is common practice and common sense. What do your members do?

The most common gas we are exposed to is carbon monoxide (CO), but we are also exposed to many other poisonous gases such as hydrogen cyanide (HCN). We need to be ever vigilant and cautious of these and many other gases that can cause death or severe injury. Maybe not today, but down the road they can and will. We are telling you to wash your gear right away because of cancer causing carcinogens, but what about the effect these poisons have on your lungs and internal organs? An excellent way to control or eliminate this danger to all is to use our gas meters to determine if the IDLH environment is safe for us not to be on air. The only alternative to keeping your people safe is to have a policy that no one is to remove their SCBA until they are clear of the building. Use your gas meters, be smart, safe, and sure.

Do you have the proper apparatus? When looking at past incidents, think about what apparatus was there initially and then later. Was what was first needed there first, or did members bring what they wanted or they felt should roll first? An example of this is we have a structure fire reported in a rural water area of our town or district. To me this means tankers are needed. Did the tanker follow the first out pump or did we send all our engines, rescue truck, and ladder before the tanker rolled? I know of instances this has happened because firefighters want to be on the pieces that will be involved in the firefight, not hauling water. Do you have run cards that specify what goes first, second, and later for your different response areas?

How is your apparatus equipped? Is it organized? Is the equipment that is needed cleaned and ready to go? If I open your cabinet doors what would I see? Everything organized and easy to reach, or a pile of junk that needs to be moved to get what we really need? Do you operate under the "you never know" syndrome? The YNK syndrome is where you save everything, even stuff from the 1950s, because you never know when you might need it. You just pile things into the rigs cabinets with no thought to access or need. Here's a novel thought: If you haven't used it in 70 years, I doubt you will ever need it again.

Many times after a run, members might be in a hurry to return to what they were doing before the call came in. It is critical that we ensure everything is put back and ready to go for the next call, before the apparatus is truly placed back in service. How often is everything inspected? Weekly? Monthly? Is it documented? Are the SCBAs and spare cylinders full or did they go back into the rig after a quick CO call, down "only" 10 or 15%? This is unacceptable. Only down 10% can equal a few extra minutes of breathing for a firefighter in a Mayday situation. Do you perform pump checks weekly, check and operate all valves, and make sure the correct nozzles are put back where they belong? After a call, is the cab's interior cleaned out and washed to remove any soot and carcinogens that might be exposing you to future dangers? In Chapter 1 we discuss washing our gear after a call to reduce cancer risk, and along those same intelligent lines we need to also wash and clean or apparatus interiors and our SCBAs.

When we purchase and outfit apparatus, we do so with a purpose and that is to meet the needs of the calls it will respond to and to ensure it is properly equipped to meet the needs of the firefighters responding. Typically, we go well beyond the NFPA standard with the loose equipment purchased. A new rig for so many means a chance to load up the bid spec with equipment that we need or desire. We need to know that the response apparatus is the correct vehicle for the call and assignment. This also applies to mutual aid responses, both going to and receiving. What you have for equipment and what you bring to a call, including mutual aid, is vitally important. If you are called for a tanker and yours is not available, would you send a ladder truck instead? I would hope not! Yet some are still making unwise decisions in the heat of the moment. These decisions might be derived from the fact you don't want to miss going to a call, you won't get in on the action, or if you don't respond to the mutual aid call with something, they might not call you again.

A sad but true example of this is found in NIOSH Line of Duty Report F2015-20.[3] I do not know why certain decisions were made, nor will I speculate. I do know from the tragic results that the decisions made increased the danger and risk to the responding firefighters. I want to share a brief overview of the incident and bring to your attention a few of the NIOSH recommendations from the report.

This department was called for mutual aid to a structure fire in an adjoining town. At the time the department was participating in a holiday festival with most of their apparatus and personnel. The only thing left in the station was a utility/brush truck, with no water and no SCBAs. It clearly was not set up for a structural response. They were called to be the FAST (RIT) team. An assistant chief, a lieutenant, and an exterior firefighter responded to the station and took that unit, knowing they were being called for RIT and knowing that the unit was not intended for structural firefighting. Now it is very easy to see clearly in hindsight or be a Monday morning quarterback, but we need to learn from this tragedy. I will leave it to you to research and read the full report on the NIOSH website,[4] but I will tell you that upon arrival, they were re-assigned to interior firefighting. Having no SCBAs they took some from the host community's rigs. The problem was the communities used different brand SCBA and they had no cross-training, even though they did a lot of mutual aid. The person who died was a 19-year-old lieutenant. It appears he had zero experience or training with the brand of SCBA he took and used. The following are excerpt portions of that report. It is lengthy but I ask you to read it, and then ask yourself could this happen where you are? Those who believe that a SCBA is a SCBA are wrong and in this case dead wrong!

NIOSH cited the following in their report:

Recommendation #3: Fire departments should ensure that special service vehicles are equipped with the appropriate equipment as specified in NFPA 1901, Standard for Automotive Fire Apparatus.[5]

Discussion: NFPA 1901, *Standard for Automotive Fire Apparatus* specifies the minimum equipment and tools to be carried on automotive fire apparatus. Chapter 6 "Initial Attack Fire Apparatus,"

Chapter 8 "Aerial Fire Apparatus," Chapter 9 "Quint Fire Apparatus," and Chapter 10 "Special Service Fire Apparatus" all require that at least one self-contained breathing apparatus (SCBA) be provided for each assigned seating position on the apparatus [NFPA 2016]. Fire departments should ensure that all in-service apparatus to be dispatched to an emergency situation are equipped with at least the minimum requirements specified in NFPA 1901.

In this incident, Department 41 members were participating in a community holiday function when they were dispatched. Fire department apparatus Engine 4110 and Rescue 4120 were located at the community function at the time of the dispatch. Rescue 4120 was used by Department 41 as a rescue vehicle and brush truck. It does not carry its own water supply. Rescue 4120 was not equipped with SCBA. Rescue 4120 was the first fire apparatus to arrive at the scene of the structure fire, followed by the Department 19 chief and Department 19's Engine 1910. Two of the Department 41 firefighters (including the Department 41 lieutenant) joined with the Department 19 lieutenant who had arrived on Engine 1910 to form the entry team. Since Rescue 4120 was not equipped with SCBA, the two Department 41 firefighters took SCBA from Engine 1910. The Department 41 lieutenant was not trained or fit-tested on the model of SCBA used by Department 19.

Recommendation #4: Fire departments should ensure that firefighters wear a full array of turnout clothing and personal protective equipment appropriate for the assigned task while participating in fire suppression and overhaul activities.

During this incident, the Department 41 lieutenant inhaled super-heated gases and products of combustion, which led to his death. It was unclear as to whether or not he had on his facepiece when he entered the basement. Documented burn injuries to his nose, forehead, and left ear were consistent with wearing a fire helmet but not wearing a facepiece or protective hood. The lieutenant also did not activate his integrated PASS device, possibly due to unfamiliarity with the SCBA he was using. The SCBA used by the lieutenant was not positively identified so NIOSH was unable to test an SCBA as part of this investigation.

Recommendation #5: Fire departments should ensure that firefighters are properly trained with the specific SCBA that they are using and also in SCBA repetitive skills training and out-of-air SCBA emergencies.

Discussion: Muscle memory, repetitive skill coordination in the operational use of specific SCBA is a learned skill that firefighters must master for each manufacturer/model of SCBA they are trained to use. Firefighters should never enter an IDLH atmosphere using an SCBA that they are not trained on, fit-tested for, and thoroughly familiar with.

Firefighters need to understand and be thoroughly familiar with the specific SCBA model that they are using. It is critically important when a department changes manufacturer or model that they provide extensive time and experience in training with the new model. If firefighters have "muscle memory, repetitive skills training" based on the manufacturer's operational instructions, they would be more able to overcome an out-of-air emergency involving their SCBA. In the aviation industry, this skill building is sometimes referred to as cockpit time. Although a pilot may have extensive experience in one aircraft, he/she needs to have sufficient cockpit time in the plane that they are presently flying in order to overcome and control an unanticipated issue. In the same way, a firefighter must have sufficient cockpit time with their SCBA because they operate in an IDLH environment and there is little time to react so the responses have to be learned and automatic. Although the principles of different SCBA manufacturers and models may be the same, the controls, visual and audio signals, and the valves and their locations are different for all models.

Repetitive skills training with SCBA is vital for firefighters working inside an IDLH atmosphere. SCBA skills training is an ongoing process that should be performed regularly to ensure that firefighters know their SCBA. The benefits of repetitive skills training with SCBA are an increased comfort and competency level, decreased anxiety, lower air consumption, increased awareness of the user's air level (noticing and using the heads-up display [HUD]), and an automatic muscle memory response of the vital function controls, such as the don/doff buttons, main air valve, emergency bypass operating valve, and auxiliary air connections (i.e., rapid intervention crew/universal air connection [RIC/UAC] and the buddy breather connection). Repetitive skills training can also provide the user with an increased ability to operate these functions and controls in a high-anxiety moment or an emergency. Many times these skills will be necessary with gloved hands, limited vision, and reduced ability to hear commands from others. Performed in conditions that are non-IDLH, repetitive skills training helps build the firefighters' muscle memory skills so their hands will be able to activate the controls with gloves on and the operation will be a conditioned or second-nature response. Firefighters have died in IDLH conditions because they did not react properly to an out-of-air emergency [NIOSH 2011, 2012b]. In this incident, the SCBA used by the Department 41 lieutenant was never positively identified.

Overcoming out-of-air emergencies is an important goal of repetitive skills training. Firefighters also need to understand the psychological and physiological effects of the extreme level of stress encountered when they run low on air or become lost, disoriented, injured, or trapped during rapid fire progression. Most fire training curricula do not include discussion of the psychological and physiological effects of extreme stress, such as encountered in an imminently life-threatening situation, nor do they address key survival skills necessary for effective response. Understanding the psychology and physiology involved is an essential step in developing appropriate responses to life-threatening situations. Reaction to the extreme stress of a life-threatening situation, such as being trapped, can result in sensory distortions and decreased cognitive processing capability.[6] In the book *Stress and Performance in Diving,* the author notes that while all training is important: "We know that under conditions of stress, particularly when rapid problem-solving is crucial, over-learning responses is essential. The properly trained individual should have learned coping behavior so well that responses become virtually automatic, requiring less stop and think performance."[7]

All SCBA come with a user's manual. Firefighters need to take the time and read these manuals, independent of the training they are given, and then practice with considerable repetitive skill building. Reading the user's manual provides only the baseline knowledge in the operational characteristics. The skill comes from considerable repetitive muscle memory training and is specific to each manufacturer and model. These skills are vital to the firefighter to assist him/her in overcoming an out-of-air emergency with the SCBA that they are wearing. Although most all of the air function principles are similar, the control locations and operations can be totally different.

The first two fire departments that responded to this incident were part of a county-wide mutual aid plan that identified a mutual aid department to respond as the FAST Team at all working structure fires. During this incident, the first firefighters on-scene arrived in Rescue 4120, which did not carry any self-contained breathing apparatus. When Engine 1910 arrived on-scene, the Department 41 firefighters grabbed SCBA from Engine 1910 that were not the same make and model as what they regularly trained on. The Department 41 second assistant chief reported to NIOSH investigators that he was familiar with the operation and use of the Department 19 SCBA. The Department 41 lieutenant was fit-tested with a size large facepiece. The SCBA that he was reported to be wearing was a different manufacturer and had a size medium facepiece. Autopsy information suggested that he did not have a facepiece on when he inhaled superheated gases and products of combustion. The SCBA used by the lieutenant was not positively identified so NIOSH was unable to test an SCBA as part of this investigation.

As I hope you can see, the observation of apparatus is a very important factor in eliminating or controlling the dangers we are exposed to. Be proactive by being properly prepared, with the correct equipment, ready to go to work. Being proactive it is the key when we talk about reducing risk to firefighters.

Water supply. So you may be wondering why I am talking about water supply in the "do I have the proper equipment" section of the chapter. Well, we need water more than anything else to put the fire out! It is, in my eyes, equipment. Look at your past operations and the water supply situation. Are you in a town, community, or district that is blessed to have a great municipal water system with hydrants everywhere or are you 100% rural with no water supply except what's in your engine's tank or your tanker (tender)? Some departments are a combination of these two. Our ability to get water on the fire is dependent on our skills to get a line to the fire and this includes establishing a water supply quickly. As you look at past operations, ask yourself how well you did with establishing the water supply. Did the pump operator start by charging the attack line with tank water? If we are in the hydrant district, was a line laid in forward lay (water to fire) or was the supply line being laid in by the next pump? Is it a reverse lay situation (fire to water)? You would be surprised by how many do not understand what you mean by "reverse lay." Do you and your members? If you are all hydrants, are your pump drivers/operators trained and drilled on drafting? Why would they need to do that? What if something happened to your municipal supply or you went mutual aid to a large fire and were assigned to draft? Would you be able to fulfill the assignment quickly and efficiently? You can't show up to an incident like that and say you are only a hydrant pumper. We all need to know how to do our jobs!

If you are a department that has to shuttle water, how quickly can you set it up and get water moving? Do you have preplanned fill sites? If you're a rural or non-hydrant district, then the fill site most likely is a draft site. How often do your driver operators drill at drafting? Do you drill with safe operations at the fill and dump sites? And the most important question: Do you get enough tankers on the road right away or do you "wait and see"? Usually waiting leads to waiting until it is too late. I would rather have four or five mutual aid tankers quick responding and turn them around than to wait and watch the fire go unchecked. Delaying calling for additional equipment until you are 100% sure it is needed can add a minimum 15 to 30 minutes of waiting for more water. Not enough water means not enough water flowing to extinguish the fire, which leads to much more property loss. Standing around watching a house burn because you don't have enough water, or you used what you had and are now waiting for tankers to arrive, is not a good thing. It affects our morale and it affects your public image.

We take getting a water supply for granted many times, but it is a much more involved process than most want to admit. Look at how well you provide water supply and if you can improve it, then provide the training and protocols to make it happen.

In this chapter I challenge you to look at the question "do you have the proper equipment?" I share many different thoughts and challenge you to honestly assess your department's capabilities with this first question. With those answered, let's go back to the incident at hand and, knowing how you operate or should be operating, ask yourself, "do I have the proper equipment for this job?" Having the proper or correct equipment can and will help to reduce or eliminate the dangers you have identified. The right equipment and the training to know how to use it can and will reduce risk, and that is what managing risk is all about.

Observation 2.
Do we have the proper knowledge?

Sometimes it is difficult to talk about knowledge because we might find some weakness in ourselves or our organizations. Yes, that is difficult to hear or think about, but how can we serve our community best and reduce, control, and manage risk to ourselves if we don't closely examine ourselves periodically? I look at this as an opportunity to better oneself and one's organization. What's wrong with being the best you can be? I don't care if you are the richest and best equipped fire department or a poor rural department trying to keep its 35-year-old pumper running. You can, working with what we have learned, become the best *you* can be.

In this observation we examine your knowledge and the knowledge of your firefighters. So let's begin by asking what does the word knowledge mean?

According to the *New American Dictionary*, knowledge is "familiarity, awareness, or understanding gained through experience or study." To take it a bit further, *knowing* is possessing or showing knowledge or understanding.

To examine this question and observation we must again go back to previous incidents, examine what happened with the question of "the proper knowledge," and look at how we performed. Did we really have the proper or correct knowledge for the incident, and what could we do to improve? To answer this, let's review five questions that will help to provide answers.

Does your department have standard operating procedures or guidelines (SOPs/SOGs) and follow them? Standard operating procedures/guidelines are instructions that a fire department assembles to help the firefighters perform routine tasks and operations in an organized and consistent manner. It can and should cover all aspects of the job. The benefits to us as firefighters are that if we all follow the guidelines, things will always get done as planned and hoped for. A simple example of this might be as follows:

> *The first engine in will lay a supply line from the hydrant to the fire forward lay, and the second engine will pump the hydrant and supply the first engine.*

A more formal and structured SOG example might be:

Mendon (MA) Fire Department SOG.
(Courtesy of Chief William Kessler)
3.7 Water Supply

Purpose:
To define and standardize our basic water supply evolutions and operations.

Scope:
A very small percentage of our response district is served by municipal and/or private fire hydrants. In these areas, they are essential to our firefighting water supply.

General:

1. When pressurized municipal or private hydrants are to be used, they are to be fully opened when in use and fully closed when

---

(continued)

finished. This usually equates to approximately 15 to 18 full turns of the valve stem. A partially opened hydrant can cause underground washout of the hydrant and attached water main.

2. All hydrants used shall be fully dressed which consists of the attachment of a hydrant valve to the steamer (large) port and 2½" gate valves to all side ports. This allows maximum usage and flexibility of the hydrant.

3. If an apparatus is going to tie in to the hydrant to pump it, the apparatus operator should attempt to position the apparatus so as not to impede traffic on the adjacent roadway. This is to facilitate further emergency vehicle traffic.

4. If an apparatus is going to lay supply line to or from the hydrant, the apparatus operator should make every effort to lay the line down the same side of the street that the hydrant is located on, so as not to impede traffic on the roadway. This is to facilitate further emergency vehicle traffic.

5. When a hydrant is opened, and prior to flowing water, an open gate valve with no hose hooked to it should be opened briefly if possible. This should be done to allow the discharge of any trapped air or debris.

6. Whenever a hydrant is shut down, the hydrant operator should make an effort to ensure that the hydrant valve is fully shut off, and that the hydrant has drained. This can be aided by leaving one cap off of the hydrant until draining is complete. This is especially important in freezing weather.

7. Whenever possible, hydrants being used for active water supply to working fires should be pumped via the hydrant assist valve. Occasionally this is impossible due to some equipment malfunction. Notify the IC if this will impede water supply. If not, notify the Chief or his designee after the incident.

8. Whenever a line has been laid from a hydrant, at least one firefighter with a radio should be left at the hydrant to properly dress it as indicated above, and to coordinate water flow with the receiving apparatus. If the manpower is needed, this position may be abandoned once the water flow has been established and confirmed.

9. Any problem with the mechanics or flow of any hydrant should be reported to an officer as soon as possible. If this occurs during operations, use of a different hydrant or an alternate water source may be required.

10. Special consideration should be given when drafting from a dry hydrant. The dry hydrant may require flushing before an adequate water supply can be obtained.
11. The IC should contact the Hopedale Water Department person if hydrants are going to be used for an extended period of time.

Developed: May 1, 2019
Revision: N/A
Reviewed: N/A

Do you have standard operating guidelines/procedures (SOPs/SOGs)? Let's not get hung up with the philosophical discussion on the terms procedures or guidelines! I have always been told that procedure means this is the way to do it, period. You have zero flexibility. A guideline is a statement of general policy. So, back to the question of do you have them. Are they written down so that all might be able to read and understand them, or is it just word of mouth? If you have them in either form, are they followed? Are they reviewed and discussed with the members at drills and department meetings, or do you just leave it to everyone to interpret as they want? SOGs are designed to keep everyone working as one. Think of them as a football play. Everyone knows what everyone is going to do and what is expected of them. If you operate without SOGs (a playbook) then how do you control the chaos? Fair question, because without a resemblance of organization and teamwork, chaos is what you will get. When reviewing past incidents, note whether SOGs were followed and, if you do not have any, would they have helped. If you are observing a current incident, are the SOGs being followed? If not, do something about it. Failure to make corrections and adjustments is a fireground failure.

If used and followed by everyone, SOGs help any emergency scene. They make everyone's job easier and make sure all are knowledgeable with what is expected to happen.

Does your department have preplans? Why preplans? Regardless of the size of your department, preplans are very useful tools. Where SOGs advise us about what we are doing, preplans tell us about what we are responding to. I'm sure that there are many volunteer departments that have extensive and involved preplans and I salute and admire them.

Preplans, even in their most simplest state, can help to identify your community by streets or sections. They identify hydrant locations or nearest water supply, type of construction found in that neighborhood, access and egress for tankers, fill sites for tankers, and utilities found such as liquefied natural gas (LNG) lines. For targeted areas or structures, they help you to identify multi-family, local hazards, commercial occupancy, and what is being made or stored inside. It might be a nursing home or senior housing, and knowledge of this will assist you in knowing about fire alarm panels and location, built-in protections, possible standpipes. With nursing homes and senior housing, your preplans help to identify what type of help we can get to evacuate and shelter those patients or people who live there. Think about it—right now if you responded to a nursing home, do you have plans in place, agencies, or companies to call that would provide assistance in helping those affected? Creating preplans and then reviewing them with the membership like SOGs will help with communicating and sharing knowledge.

Think as you review past incidents if preplans helped or would have helped. At a current incident it is about supplying information. Preplans can and will give you the pertinent information you need.

Preplans help. Remember that when things go bad is not the time to begin thinking about how to handle the situation.

What is the training and experience level of your firefighters? Training is
what makes or breaks a fire department! Later in this book we discuss training in more depth, but for now let's evaluate past incidents and ask, what level of training do our firefighters possess? Think back and honestly evaluate how different calls went. If your benchmark is "if the fire goes out (eventually), then they have all the training they need," you might be heading into a future problem or tragedy.

Nationally, we have as a training baseline NFPA 1001, *Standard for Fire Fighter Professional Qualifications.*[8] It breaks our level of training into two parts: Firefighter I and Firefighter II. Each level requires a certain level of knowledge and skill. This knowledge and these skills are intended to help us do the job safely, efficiently, and effectively, working as a team. In essence this creates a baseline for all firefighters to achieve. When we learn common skills we are able to take firefighters from different departments, form a company, and assign them a job. This is a common occurrence in the volunteer fire service. It might be raising a ladder, advancing a line, or conducting a primary search. If we all have the same level of training and the same skill set, this can happen. If we don't, bad things can happen. A decade ago, I had an old time chief tell me that everything his firefighters needed to know he could teach them in twenty-five hours! Frankly, this statement and attitude scared the life out of me! Firefighters today need skills and knowledge to do the job and work safely. We are talking hundreds of hours of training, not twenty-five.

As volunteers we want to be accepted as firefighters by all. Yet so many of us are willing to discount our skills and knowledge level because "we are only volunteers." This sounds like an excuse from the 1950s and it is not acceptable now, regardless of where you are a volunteer! How can you think we are all the same when skills might be lacking? Fires are dangerous and our job is to put those fires out! How can you think we can have a reduced level of knowledge and skills? Earlier in this book I stated "we are not all the same" and that is true, but basic skills for all must be our foundations.

What level of training do your members have? Are they trained to the level of Firefighter I/II? Are they certified to that level? If not, what training do they have to deal with the incidents they respond to? Firefighting is a very dangerous job, firefighters get injured or die doing our job. There is only one thing that is will make your members safe and capable and that is proper training.

Always remember, regardless of how much bluster and bravado you and your members have, if I look at an incident you are operating on, your training or the lack of training will show very quickly. Training is not something poor departments can say they cannot afford. Training is not something larger volunteer departments can say they are the best and are all set. Quality training is for all of us.

With all this in mind, look at the incident you are currently operating and ask if the members' training meets the needs of the incident. I assure you it will be obvious to you and all.

What about experience? Do you or the other members operating have any experience with the type of incident before you? If so, how did that incident go? How is it going now? We should be learning from our past experiences, good or bad. They will help us to see what our strengths are and where we need to improve. Past experience will help you to focus training where it needs to be.

Are the officers properly trained to lead the company and deal with the incident? What are the roles and responsibilities of your company officers? I think a big part of the job is that they are the bridge between the incident commander or chief and the firefighters. The IC gives the company officers tactical assignments (for example, primary search, first floor), and then the COs takes this assignment and break it down into tasks for the crew members. The crew is the horsepower doing the job with the officer leading and directing so as to meet the objective and assignment given by the IC. Their number one job with the company is safety and getting them out alive! They also have other jobs, perhaps being responsible for a certain piece of apparatus or a still alarm group. This is a lot of responsibility. So, what training have they had that prepares them for these important positions? Have they had formal leadership and management training? Do they have the firefighter skills to do it, not just say it? This is an important statement in the volunteer fire service. In its entirety it goes, "You might be able to say it, but if you can't show it, you don't know it!" As volunteers we all, including the officers assigned to lead a company, must possess and be able to do the skills that are needed for the assignment. This is leadership in its most basic form. I have what I call my "ten foundation stones of knowledge" that officers need to possess:

1. Effective leadership
2. Understanding fire behavior and modern fire science
3. The ability to read and understand smoke
4. Decision-making and incident action planning skills
5. Understanding strategies, tactics, and tasks
6. The ability to give a brief initial report
7. How to conduct a size-up, initial and on-going
8. Understanding building construction for the fire service
9. Know how to manage risk
10. The ability to maintain accountability and crew integrity

What training and capabilities do you and your officers have in these disciplines? What officer training have you taken or has been offered to your officers? Today's officers need a lot more than five years in the department as on-the-job-training. Your decisions and actions matter and truly can be a matter of life and death. A very good and fair question I like to ask is "Do they use common sense?"

What are the prerequisites or required qualifications for your department's officers? No matter how small your department is, this is a fair and important question. Length of service is truly a sign of dedication, but can they do the job of leading a company safely and effectively? Regardless of whether you elect or have another promotional process, I urge you to have training and skills qualifications for those who will lead. They not only lead, they in many ways control the life and death of the company members when they enter a burning building.

I once was told that the fire truck's officer was responsible for the care and maintenance of the rig and its equipment. That makes sense and you should have a person capable of doing that. Now, what about the most important equipment on the rig: the firefighters? Is the apparatus officer capable of leading them into a burning building, capable of making tactical decisions that might affect the lives of the company, and can the officer, if needed, establish command and run the operation? If not, then perhaps you need to rethink the role. In the Navy, a fighter jet has a pilot. In effect he is the company officer carrying out tactical decisions. The aircraft is maintained and watched over by non-flying personnel led by a crew chief, however. In essence it is the crew chief's plane, which he "lends" to the fighter pilot. If your new officer is not very mechanical so as to maintain the apparatus consider appointing a "crew chief" who works for the officer but is responsible for the care and maintenance. Perhaps this example might help you to make sure both the equipment and the members are taken care of.

As volunteers we know that each of us has greater strengths and capabilities in certain skills than some of the other members. This needs to be considered at each incident. Play to a person's strengths and you will get better results.

As you look at the incident before you and prepare to make tactical assignments, you must assess the capabilities of the officers or acting officers. Assigning the right task to the right officer or acting officer and company will help you control or limit the risks they are exposed to.

Is the incident command system (ICS) being used and do all know how to use it and work within it? As I began to write this section I thought of an NIOSH LODD report about a volunteer assistant chief who perished in a collapse. The first listed key recommendation was:

> *Fire departments should ensure that a single, effective incident management system is established with a single, designated incident commander, especially when multiple fire departments respond together.*

I encourage you all to go online and read NIOSH LODD report 2014-18, *Volunteer Assistant Chief Killed and One Fire Fighter Injured by Roof Collapse in a Commercial Storage Building—Indiana.*[9] It brings to light a lot of what we are talking about.

Are you using the ICS system right now? A lack of consistency in using the ICS system has been cited numerous times in NIOSH line of duty death reports. The lack of incident command is number two of the The National Institute for Occupational Safety and Health (NIOSH) top 5 casual factors of firefighter deaths and injuries on the fireground.[10] It's ranked for a good reason—failure to use it has been a factor in firefighters' deaths and injuries.

Using the ICS gives us all the ability to work as one, together, for the common goal. It helps clearly define who is in charge of what and it gives the IC a great organizational management template to work within. How does your organization run? Do you follow ICS, or have you revised it, changing titles, functions, and more because "that's the way we do it here"? Remember the purpose of the ICS is to give us common terminology and a common command template. I remember some years ago a fire department I knew wanted to simplify the ICS system. They created their own terms and structure. I was told it was to simplify the system. Well, three things happened: First, they had to re-educate their members who had learned the ICS system as it is with their new system, and second, they discovered that mutual aid was not educated in their "simplified" language. Finally and third they defeated the purpose of the ICS system and made it more complex for others to use because "that's the way we do it here"!

Yes, there are things in ICS I believe are confusing or not a natural flow for us. For example, let's say that you as the IC decide to delegate the operational aspect of the incident to a different chief. Most of us would call that person the *operations* (or OPs) *chief*, yet a true ICS purist will say that person is the *operations supervisor*. The word supervisor is a word most of us have never used on the fireground. A small tune-up like this is fine and most likely will not even be noticed by the ICS gods.

Some years ago, I worked a general alarm fire that involved over fifty different fire departments. Do you think without a common language we would have had clear, concise, and effective communications? Most likely not. The basics of ICS will work for most of your incidents. Learn it, understand and practice it, and use it at all incidents. The more you use it, even at smells and bells, the more it will feel natural at a complex operation.

To ensure that the system is used properly, we need to provide better training for all our members. During those times when we prefer indoor drills, have an ICS tabletop training class ready to go. Break the members into groups, companies, or other units. Give them tactical assignments exactly as if you are talking on the radio, and have them answer you in the same way. Break it up, ask for updates and benchmarks. Call a Mayday! Make it realistic. Sitting around just talking about ICS, or worst yet, reading slides to them, is not great training. Teach them, then have them do it.

Using the ICS keeps you organized, the chain of command clear, and everyone in full understanding and talking the same language.

Observation 3.
Are our personnel adequately protected?

Protecting personal is a lot more than making sure they wear their PPE. In this segment, protecting our people means avoiding risk if possible. Let's look at eight different ways you can work toward assuring your personnel are adequately protected:

- Identify the flow-path before committing people into the building and know the science of fire and the dangers we are exposed to.
- Venting needs to be coordinated through the IC.
- Know your building construction. Try to identify if lightweight and/ or truss construction is involved.
- Ensure your officers maintain crew integrity and always use an accountability system that works.
- Use the correct line size and nozzle configuration to ensure that you have adequate gpm flowing.

- When assigning a tactics, make sure the company is capable of the assignment.
- Have a rapid intervention crew ready and on scene.
- Monitor the time the building has been burning and how it relates to the type of building construction.

As we review each of these points, think about your training, knowledge, and capabilities; the training your firefighters have; and the operational habits of your department. Some of this overlaps on subjects we have already discussed, but they are important to look at again in the context of are your personnel adequately protected.

Identify the flow-path and know the dangers. In Chapter 3 we discuss the new science of fire. Unless you have been trapped in a cave these past 10 years or so (as of this writing), I'm sure you have heard the terms like flow-path, vent limited, and resetting the fire. Do you buy in? If so, do you really understand it? If you don't buy in, I suspect you really don't understand or don't believe in learning new things. The new science of fire is reality. It is not that difficult to understand and, based upon modern construction and number and type of modern furnishings, it makes a lot of sense. I have found from years of teaching experience that an understanding of fire behavior is really lacking in our volunteer service. This is not a putdown but rather a call to action.

As the incident commander arriving first due or later into the incident, you must identify the flow-path(s) and this is an ongoing process. Your number one job is always life safety—to protect your people—and you cannot do this if you do not understand or don't buy in (or whatever excuse you might have). Reduce the risk, learn, and use this knowledge. Lives depend on you knowing it.

Coordinate venting. With the recognition of flow-path and modern fire science, uncoordinated ventilation has become even more of a concern than ever before. Yet, we can go to YouTube and watch endless uncoordinated (freelancing) venting of fires that drew the fire into areas previously uninvolved. If firefighters had been in those rooms or areas the fire was drawn to, we could have had another LODD funeral. In the days of past practices, we thought nothing of opening up (taking as much glass as possible). It was what we were taught and, at the time, it made sense, as it was thought to expose the fire and lift the smoke to make it more tenable for us and for any civilians trapped. With today's fire load and the intense fires that come with that fire load, along with homes that are sealed up tight for energy efficiency, it is a totally different ballgame. We have always been told that venting is the systematic removal of heat and toxic gases. The keyword then (and now) is *systematic*, as in coordinated by the person in charge.

With the flow-path identified, the IC can select and coordinate where he or she wants the ventilation opening located. Ventilation will create an additional flow-path, and it must be done with thought as to where the firefighters will make entry. Failure to understand this, and failure to coordinate the ventilation assignment, can and has killed and injured firefighters.

Know your building construction. In Chapter 3 we talk about building construction. To drive that message home I need to say that as an IC, company officer, incident safety officer, and frankly as a firefighter, you must understand building construction. You need to know how it is built, the way it will burn, and how it will collapse. For most of us, volunteers and on-call, the majority of our fires are in residences and most likely Type V buildings: wood frame. Within that category are post and beam, balloon, and platform construction. Within platform we now have lightweight construction including wood trusses, I-beams, modular, and our newest concern: podium or toothpick construction. You are charged with protecting your people and reducing risk. How can you protect your firefighters if you don't understand the dangers you are exposing them to by putting them on a truss roof or an engineered lumber (I-beam) floor above a basement fire? We need to identify where we have lightweight wood construction. This can be done in many ways. For example, if your community or state requires building permits, you might have an opportunity to review those plans. If not, most of us know where there is new construction going on, so go look. Read and study this subject, understand all the classifications, and be prepared to evaluate the burning building's construction type and dangers in order to limit the risk to the firefighters actively operating.

Ensure officers maintain crew integrity and use an accountability system. *Accountability.* The subject of accountability on an emergency scene is one that needs a lot more conversation. When teaching, I usually ask the class these three questions:

- How many of you have an accountability system?
- How many of you can say it is *always* used?
- How many of you can tell me that if the need arises you can quickly determine who is where, doing what, and with whom?

The most common answers (and I bet you can guess!) are: yes, no, and no.

For me this is frightening and concerning, and I hope for you as well. A *lack of accountability* is the third most common NIOSH LODD factor cited. Being accountable is easy to do if we keep it simple, enforce it at all incidents, and maintain it at the command post. For so many of us, this means change to the ways we do it here. (Yes, I keep repeating that statement and will continue to do so. I wish this attitude—"that's the way we do it here"—could be listed as a NIOSH factor as it is commonly the root cause.) If you want to be proactive in protecting your people, here are a few thoughts on an effective and workable accountability system. As you read this, keep in mind that it is not electronic, does not have an app, and batteries are not needed. It is back-to-basics old school, but it works.

On an emergency scene an *accountability system* is something that will tell us:

1. Who is on the scene (individuals)?
2. What company are they on? (Who are they working with?)
3. What is their company's assignment?
4. Where are they working?
5. Who are they working with (other companies)?
6. When are they clear of the assignment with a PAR?

This can be done with a simple system that I share next. You can tune it to meet your needs. With this system you must do the following:

- Conduct training as to why we need it and how we are going to use it.
- Enforce its use at all incidents.
- To be most effective, use it in conjunction with a command board and have an accountability officer.

There are a few different physical systems that you can use. The choice is yours, but it is the core of a successful accountability protocol.

The system requires that every member have an accountability tag. There are two easy methods to achieve this. The first is something like a livestock tag (fig. 4–1), with the member's name, company if applicable, and the fire department they are on. You can embellish it if you want. It works easiest if it is on a snap ring. The second method and a very good and popular option is the passport system (fig. 4–2).

Fig. 4–1. An accountability tag looks something like a livestock tag.

With the tag system, when responding to a call, the members take their tags and attach them to a company ring (fig. 4–3). The company officer then creates the company ring. This is easy if they are responding on the rigs. If they are arriving in personal vehicles, they must form as companies, have a designated officer or acting officer, and attach their tags to the company's ring. With the passport system, the member's personal tag is attached with Velcro to the bottom of the helmet brim. When responding, the members remove their tags, place them on the "company passport," and the officer maintains control of it (fig. 4–4).

We only want to track those in, under, on, or operating around the structure or incident. When the company is given an assignment, the officer gives the company ring or passport to the IC or the accountability officer at the command post. That person clips it to a board and writes the company's assignment on the command board (fig. 4–5).

As you add in other company assignments, you have a complete picture of who is working (company and members), where they are, what they are doing (assignment), and who they are working with (other companies).

Fig. 4–2. Another option to account for a fire department member is a passport system.

Fig. 4–3. Members can attach their tags to a ring in a tag system.

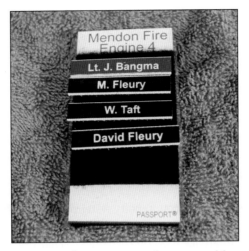

Fig. 4–4. A member's personal name tag can be removed from the helmet and placed on a passport where the officer accounts for it.

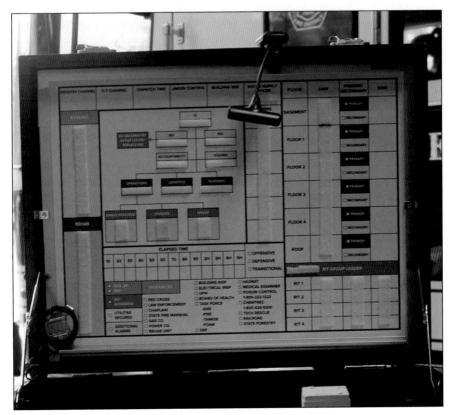

Fig. 4–5. A command board keeps track of company assignments.

When the company clears the building or assignment, they report to the CP, the officer announces he or she has PAR, collects their company ring or passport, and goes to where they have been ordered (rehab, breaking down and taking up, etc.).

This is just one example of a simple and easy-to-use system. I really don't care what system you use, just that you have one, everyone uses it every time, and if there is an emergency or Mayday, the system will quickly help determine the depth of the problem.

Having and maintaining accountability on the emergency scene assists in reducing risk and protecting your people.

Crew integrity. We have an accountability system that works, but do your officers and company members maintain crew integrity? What does *integrity* mean? Its definition is "rigid adherence to a code of values." What values are instilled into your firefighters as they relate to working as a team and in companies? For me, a few of those values are to stay together, don't freelance, and maintain situational awareness of each other. As a company, we rely on each other to do our jobs. This means you never abandon your team. We each have a role or job to fulfill and the others are counting on you accomplishing it. If you abandon them in place or decide you know better and go off doing something else, you are letting them down, increasing their risk, and possibly putting them in a very dangerous position. Worse yet, if the officer goes looking for that member and cannot find him/her, we now have a member missing and a Mayday situation. The RIT is activated and many other firefighters are now putting their lives at risk because someone, or you, decided to break the integrity of the crew. As an officer and as the chief officer of a department, you must enforce strict adherence to maintaining crew integrity! We enter as a team and we exit as a team. If you split your company leaving one member at the bottom of the stairs feeding hose up to you and your other company member(s), as long as you know where that member is and what he or she is doing, and as long as that member stays with the job, you have crew integrity.

To try and sum this up I can share a story a fire service friend told me recently about a fire in his town. The chief was out of town for the day and the deputy chief was in charge. A structure fire was toned out and upon arrival they had a large home about 30% involved. An interior attack was initiated, and more help kept arriving, both from members responding from home and by mutual aid. This friend was on-scene and overheard the deputy, who had just arrived, ask the first due captain, who was IC, who was in the building. The officer had no idea, even though it was his job to know who was on scene, who was doing what, and where they were. There was zero accountability in place. The deputy took command.

When mutual aid arrived, they tried to turn their company accountability ring into the IC and were told to "just leave it on the seat of the pumper over there." (In other words, I could care less about it.) The good news is that the fire went out, no one was hurt, and all members were accounted for. The irony here is that this department's chief is very good at maintaining accountability, and strictly adheres to it. On the other hand, the DC is not really big into it, thinks it is not really needed, and has a reputation for not using it when in charge.

This is the stuff that creates fireground injuries and deaths.

Use the correct line size and nozzle to ensure adequate gpm. Earlier in this chapter we talk about hose size and gpm. I hope that either you agree and practice the principle or that you have seen the light and are going to start practicing the concept. If initiating an interior attack, always use the correct size line and nozzle, and deliver the gpm needed. As an example, let's say we have a room and contents fire that is extending. Typically, the line of choice for this situation is going to be a 1¾" line. When using a 1¾" line we want to flow at least 150 gpm. Today, the nozzle of choice is becoming the smooth bore (back to the future again). A smooth bore tip of ⅞" will deliver 160 gpm and a ¹⁵⁄₁₆" tip will deliver 185 gpm, both at 50 psi at the tip. If you are using combination nozzles, make sure that they have a minimum flow of 150 gpm. You can identify this by looking at the stem. The number stamped is the gpm flow at the correct nozzle pressure. In the past, most combination nozzles were always 100 psi; now you can get them with operating pressures of 75 and 50 psi. Less nozzle reaction, less work. Regardless of what type you have, make sure you are getting the correct and safe flows needed.

As an aside, when looking at today's modern fire science and attack, we talk about cooling from a safe location (exterior). UL's Firefighter Safety Research Institute in conjunction with the FDNY conducted numerous evolutions on Governors Island, NY, regarding modern fire tactics. The studies that were conducted used ¹⁵⁄₁₆" smooth bore nozzles flowing 185 gpm.[11] Again remember that gpm really does matter.

Allowing your members to enter a burning structure without the ability to flow the correct, needed, gpm is asking for trouble. You are not protecting them and you are greatly increasing their risk profile. If you don't have what is needed, members should never be allowed to enter the structure.

Fireground hydraulics is a whole other subject.[12] Knowing what to properly pump all the different lines and various hose/nozzle configurations is extremely important. There are many "rules of thumb" and shortcuts out. Also, there are apps available. Regardless it is important to realize that the proper pressures are needed to achieve the required or wanted gpm's.

Now, let's look at a different problem that most of us have if we fight fires outside of a municipal water area (as in no hydrants):

> Upon arrival, you report a residential structure fire, with heavy fire showing. Experience tells you that this house is a loss as the fire has gained and is burning in about 40 to 50% of the structure and is spreading fast. As we are not near any hydrants, this is going to be a tanker operation. The tanker just signed on and is 12 to 14 minutes away, and you have a pumper with 1,000 gallons in its tank. The neighbors are gathering and this could get ugly as they are yelling at you to do something. We know that 1,000 gallons is not going to do anything. If you blitz it with a deck gun or small ground monitor you will blow the tank in a minute or two and then be standing there with no water for 10 or 15 minutes, with people yelling at you and the firefighters, a lot of tension, calls for action, and anxiety. Can you imagine what they will be saying and screaming at you? And, don't forget, they all have smart phones and most likely are already posting video on social media of the firefighters letting the house burn as they didn't bring enough water.

At this type of situation, we know that the public expects us to put water on the fire until it goes out. They refuse to understand what "outside the hydrant district or rural water supply" means to them, or you. They want to see water flowing and you trying your hardest. So, let's give them what they want. Here is what we know:

1. You don't have enough water to put the fire out.
2. The public has no idea what the correct amount of water is.
3. The public expects you to put water on the fire.
4. The tanker may have made a difference if it was here (it's not) and we initiated an aggressive attack.
5. Let's give the public the action they expect.

How? Pull your 1¾" line and pump it just enough to get a stream that does not droop but rather you can spray into the fire from the exterior. Remember, use the minimum pressure to make this happen and you will use the minimum gpms. If done correctly, you will be flowing somewhere in the 60 to 70 gpm range. Allowing time to stretch the line, get the kinks out, and flow water, you will have about 15 to 17 minutes of uninterrupted water flow. This is what the public expects: water on the fire. The public has no idea what our job is all about or how difficult it is.

They pay taxes and they want action. If you try to explain why we don't have enough water, you are wasting your breath.

I was taught this "technique" many years ago and it was then and still referred to as a public relations (PR) stream.

Situations like this are unfortunate. Part of your job is to protect your people, not only from the risks and dangers of the incident, but also to protect them from an angry community or property owner who has no clue about our job. Being ridiculed, put down, and yelled at is not good for our morale and self-esteem. Protect your people.

When assigning tactics, make sure the company is capable of the assignment.

On arrival, there is a call to action. You are expected to start making decisions and giving tactical assignments (the jobs you want them to do) and the pressure is on.

Wait. Stop and look at the make up of the companies on hand or when they arrive, and determine whether they are capable of carrying out the assignment. When you make an assignment, you are essentially assigning risk. Risk is a part of the job; it is something that goes hand in hand with what we signed up for. However, you need to limit the risk and protect your people. How? If you make a tactical assignment to a company that is not capable, you are placing them in danger and exposing them to high risk. Risk both to the members of the company and to others that might be operating on the fireground with them. Unfortunately, there are those who still think age means experience, length of service trumps capabilities, if you have the title firefighter you're all set, and to say we can't do something is to show weakness. This means it is incumbent upon the IC to make sure those assigned can do the job. If you make the assignment via radio and upon their arrival realize they are not the correct company, that they are not capable, then you must pull the assignment back and assign it to another company that is stronger and more capable. This is called *risk transfer*. It takes the high risk and reduces it by reassigning it to a company that has the capabilities; and even though you are giving the new company risk, it is calculated risk based on their capabilities and training.

A good IC must know when and why risk needs to be transferred. Remember, when we are called to action we have a job to do, and the last thing we need to worry about is hurt feelings. You know what I mean, but the bottom line is minimizing and reducing risk here. So, when it comes to hurt feelings, you need to stop and think why did you make that reassignment that caused the hurt feelings? If they were not capable, was it physically or was it skills? With lack of skills capabilities it is usually a training issue as they never show up for training, or when they do, they don't actively participate. Why? It could be the "I'm all set" attitude, the "I'm an officer and know my stuff" attitude, or, worse yet, the "I've been doing this for 30 years" attitude. Regardless, if they have poor or no skills for the job at hand, transfer the risk. As for physical capabilities, as a retired old timer, I can tell you that for my last few years of active service I found my physical capabilities fading away. I'm not ashamed of that because it's a fact of life. We all talk about it's a young man's game because it *is*, and we need to face it. Don't worry about hurt feelings because you have a huge responsibility on your shoulders. Do what is right for your operating people and don't worry about the whiners.

As part of this, you might find that the crew is the correct crew and very capable, but that the officer leading the crew is not capable. The roles, titles, and qualifications of the officers in many of our departments are sensitive subjects. You know I am an advocate for officer training and qualifications. The role of an officer making entry is to lead them in, achieve the tactical assignment if possible, and get them out alive, which is job number one. If they are not capable leaders, how does this happen? Does the officer give the instructions and send a crew in while he or she stays outside? Don't laugh; I know of many instances where this has and does happen. That is one of the reasons I am writing this book. If the officer is not the right one for the job but the crew is, reassign the crew members to a more capable leader, transfer the risk, and do what you know is right. This truly is protecting your people. They might not agree or appreciate it, but this type of action is the doing the right thing.

Risk transfer is a very important action when we ask, Are we protecting our people?

Have a rapid intervention crew. If you look at NFPA 1250, *Recommended Practice in Fire and Emergency Services Organization Risk Management*,[13] nowhere does it talk about, encourage, or discuss a rapid intervention team. Managing risk is a lot more than words, theory, and definitions. Having a RIT ready to go, if needed, is an important part of any risk management system, especially in our volunteer service. We discuss rapid intervention in much more depth, including how to staff a competent team, later in this book. For now, imagine you have a Mayday right now, and there is no RIT or the one you have is incapable. Now what? Waste time trying to find the correct people, or hope and pray the downed firefighter survives? We must remember that the rapid intervention team has one purpose and that is to rescue one of us if we become trapped, lost or injured. RIT is for us. Have a capable team ready to go at all structure fires. Protect your people, be proactive in risk management, staff a RIT.

Monitor the time the building has been burning. In this book we discuss many things that you as the person running the incident (IC) need to know and understand to be a safe and effective IC. Now we relate that with what we know about the structure and it's type of construction, where the fire is located within the structure, the contents, and how long it has been burning. Common sense tells us that the longer it burns, the more dangerous it becomes for an interior attack. Is there a basement fire? Is lightweight engineered lumber (wood I-beams) used and you are putting members on the floor above? You should be aware that the prolonged burn time will soften or melt the glue holding the engineered lumber together, and it is prone to failure quickly. Sometimes in our aggressiveness we fail to realize or remember that the longer it is burning, the more dangerous it becomes. Know your construction and apply that knowledge to better protect your people. As mentioned earlier when we talked about preplans, if you have buildings going up in your community, make the time and effort to research the types of construction—especially if lightweight construction is involved. Time does matter. Here are a few ways you can track the time into the incident:

- Have a digital stopwatch on your command board. When you take the board out, start the timer.
- If you have quality dispatching, have them monitor the time and provide benchmarks to you. For example: "Fire alarm to Main Street command, you are 20 minutes into the incident."

Remember, the longer it burns, the higher the risk. Protect your people by keeping track of prolonged burn time.

In this chapter we look at the three key areas of observation to help control or eliminate the risks to our firefighters. Conduct a review of how your department operates in conjunction with the three observation points to discover what works well and where improvement is needed. After any needed improvements are made, using this system at an emergency scene will be fast and easy to manage.

Survivability Profile: When an Incident Involves Trapped or Missing Civilians

Along with the three key observations, there is another subject that overlaps all three and needs intense risk management: the report of people trapped or missing. Many of us protect communities, especially smaller ones, where everyone seems to know everyone. The vast majority (about 70%) of volunteer departments cover populations of fewer than 5,000 people. I have been involved in three tragic

incidents (people perishing in the fire), and at two of them, members of the department, including myself, knew those who perished. It was a very difficult time for
so many of us. We as volunteers have a higher likelihood of knowing those who
call us for an emergency than our brother and sister urban firefighters. This is a
fact of life in our smaller communities and we need to address it here. Just imagine this scene:

> You're toned out for a structure fire and the address is familiar. On
> arrival, you know the house! It's your friend's elderly parents. Then,
> you are told that the father is still in the building. We know he is
> housebound and uses a wheelchair. The flames are rolling out and
> the neighbors are screaming at you to save him. The pressure is on,
> the adrenaline is pumping, you're scared, and the call for action is
> overwhelming.

A situation like this is not that farfetched for many of us. Even if you don't
know the person, people trapped or missing is not a call we go on every day. More
than ever, you need to remember that your people are looking for you to do the
right thing. In order to make the correct decisions, you can apply a survivability
profile to the person missing or trapped. To some of you, this might sound cold
and you are thinking, Hey wait a minute, I know the person who's trapped! I truly
understand. I have been there and it is not a good feeling. But what I also know
is that as the IC I am expected to make logical, hard decisions. Remember our
three fireground priorities? The first is life safety. Using a survivability profile
before committing firefighters to a dangerous rescue situation might spare your
firefighters from injury or even death.

Step back and look at the building. *Apply common sense, not emotion*. To conduct a survivability profile there are two things to look at and consider:

1. What are the fire and smoke conditions? It the entire structure burning
    or is it a room and contents fire? It the building pushing heavy, black,
    fuel-enriched smoke? What are the heat conditions for the victim?
    Will it most likely flash shortly after opening up? Remember, you're
    going to be putting firefighters in there.
2. Do you know where the victim is located? How long has the victim
    been trapped or missing? Is the person in a room that is fully involved
    with fire? If not, is the person exposed to the heavy, toxic smoke and,
    if so, for how long?

You are asking these questions because we need to establish the condition of the
victim. Without emotion, honestly ask yourself is the person still alive or is he or

she long expired? It is one thing to aggressively commit a company for a quick rescue of a person who has a chance of survival, and another to put a company in horrible danger for a corpse that is burnt beyond recognition. I often tell my students that if you recover a viewable (at the funeral home) body for the family to have closure, that is a good and honorable thing. But if it is an unrecognizable corpse, the casket will be closed anyway, and that might not be worth endangering your people. Committing a crew in a high-risk situation to save a long-expired civilian is questionable at best.

Now if your survivability profile survey tells you we have a good shot at rescuing the victim, then you have to look at something else before committing to the rescue, and that is the on-scene crew's capabilities to attempt a rescue and get back out alive. Remember, we are not all the same. Our skills and physical capabilities vary by member. In a rescue situation, time really matters. We cannot wait for other, more capable firefighters to arrive or respond through mutual aid. Someone is dying and needs our help. You as the incident commander must look at your on-scene staff and honestly ask if they are capable to attempting this rescue. If not, you cannot commit them. You may have a group of old timers who brought equipment or responded to help out. Or you might look around and see a bunch of members who really don't have strong firefighter skills. Yes, we usually have a job for anyone who wants to volunteer, *but right now you need trained and capable firefighters*. Attempting a search and rescue is a matter of training and skills. Remember all those drills where the attendance was disappointing? Have you ever heard or been told "I'm all set, Chief, I've done the search drill before." (You mean 10 years ago?) Your members' training and skills really matter, and in a situation of people trapped, they matter right now, especially to the civilian trapped and dying.

If you ever have a situation as described here, remember that emotions can and will cloud your judgment. If it is too difficult for you, enlist someone to help. It might be a family member or close friend who is trapped. Asking for help, as in transferring command, is an honorable and smart thing to do in cases like this.

Your number one job is to control, eliminate, or avoid risk. Protect your people. Read, study, take classes, and keep your abilities sharp and accurate.

Managing Risk in the Volunteer Fire Department

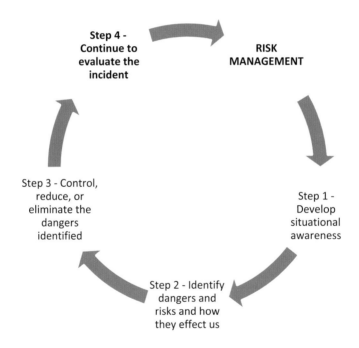

In the previous chapters we review the first three steps of our components of a risk management system: developing situational awareness of the incident, understanding the dangers and risks, and determining how to control or eliminate those dangers to reduce the risks. We also review and talk about the questions that need to be asked as they relate to having the proper equipment and knowledge, plus the question of are our personnel adequately protected, and discuss the importance of conducting a survivability profile before committing members into an intense rescue situation.

With all the questions answered, and the dangers, risks, and actions taken to reduce the dangers and identify risks, we now have a working system and have gained knowledge. It is now time to make sure we keep it working as intended.

Step 4. Maintain an Ongoing Evaluation of the Incident and What Is Happening

We know that from the very beginning of an incident we must constantly evaluate what is happening. For many of us, we have been taught this as it regards size-up. The size-up starts with a brief initial report and then continues with a complete and thorough size-up, including a 360-degree look. The size-up must continue throughout the incident until the last firefighter clears the scene. This is the same size-up we have been doing for a long time that helps in maintaining situational awareness. It is also a key part of determining if the incident action plan (IAP) is working and if we have the correct personnel and equipment on-scene. Size-up is a key component of any risk management process.

To begin the ongoing evaluation step, first ask, Is my incident action plan working? You base your plan of action on what you discover and analyze when you arrive. Now that you are into the incident, you must determine if your plan is actually working or if you need to adjust it.

To determine if it is working there are some basic questions to ask:

1. Are the companies making progress and achieving their tactical assignments?
2. What has changed since arrival and how is it affecting the operation?
3. Are all the companies staying with the plan or is freelancing beginning to happen?
4. Are you as the IC getting updates, progress reports, or benchmarks?
5. What is the physical condition of your crews operating?
6. Is the weather adversely affecting the operation?
7. Has rehab been established and are your members rotating through it?

Are the companies making progress and achieving their tactical assignments?

What have you expected to happen? Is it being achieved?

What has changed since arrival and how is it affecting the operation?

Is the fire growing unchecked? Are we able to suppress the fire? What is happening with the smoke? Remember, being able to read the smoke gives you an excellent indicator of what might actually be going on. Are more companies arriving, allowing you to make other needed tactical assignments? How is your water situation? Is it adequate and in place or are you "working on it"?

The easiest way to sort this out is to honestly ask yourself if the situation is getting better or worse. If conditions are improving, we have the fire in check, and companies are making progress, then the incident is improving. On the other hand, if things are not improving you will need to take corrective actions. Many times we find that when things are not going well the radio traffic will be a sure bet indicator. Voices rising in pitch and tone, people yelling and talking over each other so no one can get a clear message through. This might sound familiar. Earlier in this book we discuss radios and communications. This is where strong procedures and protocols can make a huge difference. If your radio traffic is out of control, it is highly likely that things are not going well and no matter what you may think, you are not in total control of the incident.

Are all the companies staying with the plan or is freelancing beginning to happen?

Within the first month I had joined the volunteer fire service, one of the guys told me to never freelance. He explained to me what it was and how working outside the system, thinking I knew better and just doing what I thought needed to be done, is *freelancing*, and is dangerous to all. I listened, and can honestly say that in my 37 years of active service, I never freelanced. If I felt something needed to happen, I always asked my company officer. And later, as an officer, I would consider prompting the IC or asking if I could take that action. (A great example of freelancing is again, ventilation.) Here we are all these years later, and freelancing is still an issue we are talking about. We have seen firefighters killed by it, firefighters injured by it, and firefighters have near-death experiences by it! It is unofficially accepted as part of the job. We should all be repulsed by the word freelancing and those who participate in such actions. Stop and think about this: We all know it is a bad thing to do, and yet here we are today still dealing with it! As a volunteer I would suggest for your consideration a three-step program for those who freelance.

Step 1.   If a member freelances, he or she needs to brought to the chief officer and told that it was wrong and could have had negative consequences on the incident and that they will never freelance again.

Step 2.   If the same member freelances again, suspend that person for 30 or 60 days, and make sure to collect the member's gear during that time. This will hopefully teach that member that freelancing is wrong and we will not tolerate it. Furthermore, it also sends a strong message to other members that freelancing is not tolerated.

Step 3.   If a member still freelances upon returning to active duty, dismiss the person immediately before someone is killed or injured!

There is no place in our volunteer fire service for freelancers, so let's end it now.

Are you as the IC getting updates, progress reports, or benchmarks?

If your company members are not updating you, how do you know what is happening? Are your company officers trained to provide you with *pertinent feedback* and *progress reports*? We have discussed unnecessary radio traffic and why it needs to be avoided. But what about the important stuff? As the IC you need to know what is going on inside the building because contrary to popular opinion, the IC does not have X-ray vision! You can't see through walls and thus you might think you know, but unless someone tells you, you really don't know what is happening inside. So, back to those radios we seem to always talk about. Our officers need to be trained to provide feedback and progress reports to the IC, even if the reports are about lack of progress.

Feedback, progress reports, and benchmarks keep you informed and updated. The acronym CAN (conditions, actions, needs) is an excellent tool to use when making a radio report.

A few examples include:

- "We've got a lot of fire up here. Requesting a second line."
- "Engine 3 has located Ladder 1 and all members are accounted for."
- "Engine 22 is reporting that the main body of fire has been knocked down."
- "Truck 2 reporting second floor searched. All clear."

These brief messages keep you informed as to what is happening behind those walls. They assist you in assessing if your plan is working, or do you need more help or resources. They help you to see if you need to adjust the plan. Many times we are hesitant to insist upon these types of reports or our officers are reluctant to provide them. Typically, it is because the department has never done them in the past, the members don't know how to do them, or they don't understand the importance of these reports or are afraid of creating too much radio traffic. Sometimes it is you having never realized how these reports can help you as the IC.

Getting benchmark and progress reports is a key component in helping your ongoing evaluation.

What is the physical condition of your crews operating?

So many times we deal with being short-handed, we working with little water and a lot of waiting for both! Being short-handed can mean that when members go low on air, they come out, change SCBA bottles, grab a bottle of water, and go right back to work. For even the best of us, this is not something that should be a common practice. Many of us old timers remember doing what I just described. However, our total PPE consisted of cotton duck coats, tin or plastic helmets, and pull-up boots. Today, we are fully encapsulated. The houses then were not sealed tight and the contents were not the fire load we experience today. We did not experience as much heat as we have today and thus we could go in and out without the physical demands on our bodies that we experience today. What this means is that we must monitor our people for fatigue and health issues. When you are shorthanded or operating with just a few people, the firefighters will be physically overwhelmed trying to accomplish what needs to be done. We know they will do their best, and they know that if they don't do it, there is no one else available. Yet, in risk management you must take care of your people first! Buildings can be replaced, a firefighter's life cannot. If you see your crews slowing down, dragging, and otherwise looking physically spent or exhausted, it is time to rethink your plan and priorities.

Is the weather adversely affecting the operation?

In a previous chapter we talk about weather when evaluating dangers and risks. Ask yourself if the weather is having an adverse effect on the job. Ice and snow will slow down tanker operations, and sub-freezing temperatures will affect hydrants and water relay pumping. The cold and the heat will slow down both your staff and the speed of the operation. All of this needs to be evaluated.

Has rehab been established and are your members rotating through it?

For many departments, the idea of rotating might sound comical. "What do you mean, rotate through it, Joe? There are only five of us here total!" I've been there, done that, and can tell you that I was physically wiped out! Having a quick rest and one or two bottles of water does wonders. As you evaluate what is happening you must look to making sure your members are provided with the chance to rest and hydrate. Even if it is a room and contents fire with a quick knockdown, those who went in and attacked the fire will be hot and in need of hydration (fig. 5–1).

From your ongoing size-up you are now reassessing the problem (incident) before you, and with that information you can logically and informatively answer the question: Is my IAP working? If not, you must be proactive and begin to make logical adjustments, changes in tactics, and, if needed, changes in strategy. Today the word proactive is used so much that we are losing sight of its real meaning. *Proactive* is defined by the Merriam-Webster dictionary as "acting in anticipation of future problems, needs, or changes." It is further defined for our purpose as a person creating or controlling a situation by *causing something to happen* rather than responding after something has happened that we did not expect or hope for. As volunteers we must always work toward looking at the problem before us in order to identify and attempt to prevent or avoid any potential problems. Being proactive is clearly a big part of a good risk management program, and is something we need to take seriously by practicing it at all responses.

It's always very easy for someone to tell you to always be proactive, but how can you become a proactive officer or IC?

Fig. 5–1. Remember that life safety includes our own members, and this includes rehab and hydration. Are your members in need of rehab and hydration? (Photo courtesy of Jim Ahern.)

To help you to stay proactive, I offer ten tips that will assist you and work:

1. Remain calm, think clearly, act decisively, and know your limitations. This is easy to say and sometimes hard to do, especially at a larger or more intense incident! We all know that. To work at overcoming this issue, make a conscious effort to conduct yourself at all incidents in a calm, decisive manner, knowing your limitations. This way you will have built the skills needed to run the big one.

2. Always remember it takes guts, water, skills, and common sense to put out the fire. You must have all four. We typically hear this phrase with only three components: guts, water, and common sense. However, if you and your team do not have the skills needed for the incident, how can you safely put the fire out? No skills, lack of skills, and lack of knowledge has led to firefighters to getting injured or killed. Skills and knowledge matter. We discuss training and ideas to attain the skills needed later in this book.

3. For every action you take there will be a reaction. Is the situation getting better or worse? Remember Murphy's law: If something can go wrong, it will. Always keep sizing up and watching for the expected results. If they are not there, readjust your tactics.

4. Know your limitations and seek help if needed. Swallow your pride! It actually is the sign of a great leader who knows limitations but also knows where to get the help and guidance needed. Do you?

5. Know the limitations of your personnel and resources, including water if you are in an area with limited or no municipal water supply. We are not all the same, nor do we all have the same skills. Always remember this.

6. Keep an eye on the time into the incident. This is an important observation as prolonged burn time adds to the risk. The longer the fire burns, the bigger it gets.

7. Remain focused on the overall incident. You are the IC, not a company officer. If you want to be on the line, then reassess why you are a chief officer. The IC is responsible for all those working and needs to manage and keep an eye on the entire incident.

8. Always use the ICS system. It works! Use it for your benefit and the benefit of all those operating. If you insist on not using it or creating your own, then you are responsible for the bad consequences that arise from lack of fireground control and management.

9. Use and maintain a personnel accountability system. There is no wiggle room here, people's lives might depend on it! *Always* know where your

people are, what company they are working on, doing what, where they are, and with whom are they working.

10.   Get more help to the scene *before* you need it. In other words, anticipate! Many departments are still reluctant to call in advance for more help. When we take the wait and see approach, the situation can grow in size and get a lot worse. We are working for a much longer period of time with not enough help and not enough water. You can change this by anticipating the needed help and calling for it earlier. It is easier to cancel and turn them round than to beg them to speed it up!

No skills, lack of skills, and lack of knowledge has led to firefighters getting injured or killed. Skills and knowledge matter.

Indicators and Warning Signs During Your Ongoing Evaluation

There are many different observations you can make to track and evaluate how the incident is going. In this section I present some incident indicators and warning signs that I think will be helpful. As you read them, be objective; you are trying to conduct an overall evaluation of your incident. We have already discussed many of these points, but I present them to you again as warning signs to look for. Remember warning signs are just that: a warning, a caution, a prediction of the possibility of a bad event that could happen. It is up to you to pay attention and use your skills and knowledge to overcome what is happening or not happening. Some might take offence to these warning signs and indicators as they will see it as an affront to the way they do things. But my intent is to help us all realize that things are different in the fire service today; we can't live in the past. Most importantly, the dangers and risks firefighters are exposed to today are extreme. Many of us are still fighting twenty-first century fires with 1960s tactics. It is time we move forward. As you review the following points, think about how you do things now and is it different than in the past? If not, perhaps it is time to update or realign your skills and knowledge to be a safer, more progressive fire officer and incident commander.

We break down the indicators and warning signs into three classifications: incident, personnel, and incident commander.

Incident indicators and warning signs that increase risk

Incident indicators and warning signs are things that might increase risk. Again, we have discussed many of these observations in a different context, but we also need to look at and remember that they are also warning signs of possible risk. Incident indicators usually mean that the incident is in charge of the fire, not you. It might not be your fault but rather the fire had such a head start that the building is full of fire, or it might be your fault because of a lack of or poor tactical decisions. Either way, you need to be aware and take some action to reduce the risk.

The following are examples of incident indicators and warning signs:

Prolonged burn time. How long has it been burning? Since your arrival is it still burning out of control? Are the efforts to knock the fire down and control it working and successful?

Extensive structural damage. Has the fire burnt to the point there is now extensive structural damage? Remember, weakened structures collapse and fall down. Sometimes with firefighters on, in, or under them! Not a structure on earth is worth the life of one firefighter.

Flow-path not identified and you have a lack of coordinated ventilation, with convection spreading the fire throughout the structure. The new awareness of fire behavior should make identifying the flow-path automatic. Failure to do so as an incident indicator will affect the way the fire extends and spreads.

Lightweight construction is compromised. You must understand the dangers lightweight construction presents to us as firefighters. When you suspect the lightweight construction is compromised, you need to consider withdrawing your people from its vicinity as the possibility of collapse is now extreme. Lightweight construction being compromised is a significantly higher danger to us than to our urban brothers and sisters. Why? Mostly because of response time. Career staff can usually get there a lot quicker and therefore the burn exposure time to the lightweight materials is less. For volunteer staff, especially in very rural areas, our response time to the scene is much greater. This is not a putdown of our service but rather a fact. We need to respond to the station, get the rigs, and respond to the fire. All this is time where the lightweight construction is exposed to fire and heat. Beware!

The incident is too large for the initial staff and they are overwhelmed.
This is a classic warning sign and one that many ignore thinking it's business as usual. Remember, if you are reading this book you are seeking knowledge on risk management. When you arrive at a large incident with an initial staff that is just too small to handle what needs to be done quickly, but you just go about things with a "damn the torpedoes, full steam ahead" attitude, that is a big red flag, a clear warning sign that you are increasing risk without much thought or evaluation.

Multiple floors are involved. When you have multiple floors involved you need to consider many things, starting with do you have enough personnel to initiate a fire attack on multiple floors? How much fire do you have and can you even make entry? Multiple floors on fire is a difficult but not insurmountable situation. The fire might be just extending upward (remember convection is the number one way fire spreads within a structure) and you will have to have some fire control on the first floor before going to the second floor. Can we get a quick knockdown from the exterior and cool it from a safe distance, keeping it in check for a few minutes while we get crews inside? If you have on arrival or during your incident the fire spread to multiple floors unchecked, you need to quickly assess the situation and most likely withdraw your companies as you do not have control of the incident.

Personnel indicators and warning signs that increase risk

Personnel indicators and warning signs are ones you can easily identify, as they usually reflect on the quality of your personnel. This subject is difficult to discuss as it directly reflects on you and your members and their abilities and skills. Be objective here. If things need to improve, be a change agent and do something about it.

Following are examples of personnel indicators and warning signs:

Not enough help. On-scene members are fatigued and have to keep going.
Again, this goes to the issue of not having enough help, at least early on. In cases like this, our members want to make the best effort they can, sometimes to their own physical detriment. Remember that we all have skills limits and physical limits. When short-handed, the IC must prioritize what needs to be done and *what can be done with who and what is on scene at that moment.* The decision should always be based upon life safety as the first priority—ours first.

Companies not maintaining crew integrity. One of the biggest dangers we face at any emergency scene is the loss of crew integrity, which then means a loss of accountability. *Integrity* is defined in the American Heritage dictionary as "1. Rigid adherence to a code of values. 2. Completeness, unity." For us as firefighters, this means we work as a team and we stay as a team by maintaining either sight, touch, or voice contact. Voice contact can also be by radio if you split the company. Another example would be to leave a member at the base of the stairs feeding hose. If the member stays at the assigned location, and you can shout out to or reach the member by radio, you have accountability and crew integrity. When crews lose their integrity it is a warning sign, a huge indicator that the IC might be losing control of the companies working and that the risks to all are increasing. As mentioned in a previous chapter, the loss of crew integrity and loss or lack of accountability is a top factor in line of duty deaths.

Lack of training. This is a huge personnel indicator and warning sign and one that is usually very easy to spot. Sometimes the IC observes this with mutual aid departments working on the fireground. Regardless of whether it is your own department members or a mutual aid company, lack of training is very evident on a fireground and emergency scene. It is a major warning sign that will increase risk. Typically, you will see people running around, in total panic, pulling all sorts of lines and equipment off the rigs but doing little else. You will also see the lack of training as the people cannot achieve the assignments and tactics you assigned. All in all, the fire will eventually go out; usually a total burn out, or as we used to say, cellar hole save. Lack of training might be members not showing up for a training event as they think they are all set, or it might be the department has a very poor and weak training program because, that is the way we have always done it. Regardless of how it happens, recognize the dangerous risk factors when you watch poorly trained people on your fireground.

The weather is affecting the crews. The final example of personnel indicators is weather. Weather certainly is not the members' fault, but it does take its toll on their performance. We all realize this, but how many of us proactively do something about it? Think about those extremely hot and humid days. Do you wait until you are on scene and confirming that there is a fire before you call for mutual aid and additional help? Doing so increases the dangers and risk to the members on scene. How so? They will have to work in adverse conditions, pushing themselves a lot longer until that help you just called arrives. This might be 15 to 25 minutes, or maybe more. What about the few firefighters you have on scene? Are you going to expect them to work hard and keep going, knowing the weather is taking a significant toll on them? On those hot days, do you the IC wear full turnout gear? A few years ago an old time chief, recognizing the heat stress full modern PPE puts on our members, told me that he always, even while standing at the command post, wears a full set of gear. His reasoning was excellent. How else would he know how the heat was affecting his members? I know of a "old-time" chief in the northeast who is known to wear Hawaiian shirts and shorts on scene on those dreadfully hot, humid days. It has become his signature and many think it is funny! Tell me, if he is dressed to stay cool, like he is going to a picnic, how does he know what his members are going through? In cases like this, save the signature clothing for the department's summer cookout and start appreciating what your members are experiencing by feeling the same conditions. The same goes for the cold, snow, and ice. Weather affects our abilities to perform and thus can increase our risks and dangers. Pay attention to it. Be proactive and anticipate more help and take care of your members on scene.

Indicators and warning signs of the incident commander losing control of the incident

Incident commander indicators are the hard one to discuss here because I am telling you to examine what *you* are doing and how *your* decisions are affecting the safety and risks. It is also an examination of your accepted practices. I am in no way trying to insult or critique you and other chief and senior officers. As the IC you must realize that what you do has consequences. Plus, we are all human beings and I have yet to meet an infallible human. Let's face it—we all have made poor or bad decisions. The following indicators will help you to correct or avoid those bad decisions going forward. The world's greatest leaders all realize that mistakes and misjudgments happen, but they are wise enough and smart enough to realize it and then they fix it. Always remember, in an IC's case a misjudgment might kill someone.

As we review these IC indicators, please keep in mind the NIOSH top five cited contributing factors of firefighter death and injury:

1. Improper risk assessment.

2. Lack of incident command.

3. Lack of accountability.

4. Inadequate communications.

5. Lack of SOPs or failure to follow established SOPs.

These command issues lie at the feet of the department's management; however, they can be easily overcome with common sense, a great training program and the acceptance that things have and will continue to change and in the fire service we need to keep current.

Following are examples of IC indicators and warning signs:

Your strategies and tactics do not meet the needs of the incident, or were never identified. As the IC, your job is to determine what your strategies are for the incident and then implement tactics to achieve them. For those who might need a bit of a refresher, simply put, a *strategy* is where you want to go, what you want to achieve. At a structure fire, a strategy could be as simple as put the fire out. *Tactics* are how you are going to get there, how you are going to do it. A tactical, or tactical assignment, could be "Engine one, fire attack, first floor." Always remember that telling me *how* to do it is a task, and task assignments are left to the company officer.

During your ongoing size-up, you need to objectively look at whether or not your plan is working. Have you done your 360? Are you making assignments to companies that keep us all working together for a common goal, or is it a free-for-all? Thinking all is going well without really determining if your plan is in fact working is a very dangerous game to play. You are increasing risk to all and it is happening without you and others even realizing it. When things are not going as planned, you need to adjust the plan or you will lose control of the incident. As the IC you must remember that you are responsible for the overall plan, the big picture. Focus on it as a coach, not a quarterback. Even the best quarterbacks report to the coach, and that, my friends, is what you are and what is expected of you. Figure it out, set up and implement a plan, and then give the players their assignments. If it is not working, fix it.

Ineffective or out-of-control communications. Our ultimate goal should be to always have great communications. We have discussed how a breakdown in communications can cause havoc on the emergency scene, and as such we need to recognize and strive toward making our communications training meet our needs and requirements. So many times at an emergency scene we hear communication transmissions that are not needed, are long-winded with too much information, or even worse, numerous people trying to talk at the same time! Have you ever tried to transmit an important message and been unable because you cannot clear the airways to send it? It's frustrating, not needed, very unprofessional, dangerous, and most of all avoidable with proper training and radio discipline. When this type of communication is happening, you as the IC have really lost control. How can you run an emergency scene when you and others cannot communicate properly and in a timely manner? The IC also needs to be sure that the transmissions and messages sent are effective. We need to make sure that those we are sending the message to understand them. The use of "10" codes and other radio shorthand is ineffective and very outdated.[1] Why?

Because they don't work and are often misunderstood. Messages need to be clear, to the point, and as brief as possible. Before you send them, make sure the party you are trying to reach is listening. And, once you have sent the message, be sure it was clearly communicated.

A simple example might go like this:

> **IC:** "Engine 1 from form Command."
>
> **E1 officer:** "Engine 1 answering command."
>
> **IC:** "E1, upon arrival stretch a line from E2 and backup their crew on the second floor."
>
> **E1:** "Officer E1, received. Stretch from E2 and back up Engine 2 on floor two."

Now some folks think that this type of communications model is a waste of breath and airwaves. Well, they are wrong. It is simple, brief, and to-the-point, plus we are ensured that the receiving party clearly understands the message.

Great communications are a necessity. Do something about it or be prepared to be in a position where you are losing or have lost control of an incident.

No personnel accountability system. This is one of the gravest mistakes an IC can make! You might go to 1,000 fires in your career and get away without having used an accountability system 999 times, but it is the 1,000th time when you need it, do not have it, and firefighters could be dying and you don't have a clue. Without a system you really don't have control! You really don't know who is on each company, where the companies are, doing what, and with what other companies. You might think you do, but if all goes wrong and Maydays start coming in, are you 100% positive you have the needed knowledge of your personnel, where they are, doing what, and with whom? Don't fool yourself! The use of an accountability system not only gives you the information you must have but using it *at all calls* sets a standard within your department that must and will be followed. It becomes second nature and that is what you want!

Your span of control is greatly exceeded. We have been taught in the incident command system that a span of control should not exceed seven companies or positions directly reporting, with five being the optimum. The more direct reports you have, the greater the demand on you for time and management. On the fireground or emergency scene you, as the IC, are responsible for the overall plan of action. But that does not mean you need to have everyone reporting directly to you. Now, if you tell me that your officers are not capable, my response is to either train or replace them. If your firefighters are not capable of completing basic tasks without you telling them how to do it, then you need to properly train them or they are not firefighters. Great ICs and fire chiefs know how to delegate and also know their staff and the capabilities of each member. When you fail to stay within the prescribed (and proven many times) span of control, you will begin to lose control. No matter who you are or what you think you are, no one can control everything and stay 100% effective. In our fire service that lack of effectiveness might cause risk and death to one of our own.

With this chapter we wrap up the formal portion of risk management as found in most texts and articles on the subject. However, I think managing risk in our volunteer fire goes well beyond a four-step process. Chapter 6 examines other actions that reduce the risk to us all.

Additional Ways We as Volunteers Can Reduce Risk

So far, we have pretty much kept to the traditional risk management agenda in talking about the four-step system and all that goes with it. Now it's time to expand our minds and our actions and our way of doing things and explore additional ways and means that we can reduce risk. None of what we're about to discuss is typically ever talked about in any risk management articles or systems, but to me they are all significant foundational stones to a successful risk management program in every department. In this chapter we talk about the incident safety officer, the rapid intervention team, and the department's training. All three are important to achieve an extraordinary risk management program and we need all three things as volunteers.

The Incident Safety Officer (ISO)

To help in the reduction of risk, always assign a highly competent incident safety officer. Unfortunately, this position has been used as an assignment for someone who is considered semi-retired, or to get him/her out of the way, or worse, because we don't know what else to do with him/her. Those days are long gone. We need to understand the ISO position and why it is important. I ask firefighters all the time: "Do you know the difference between a safety officer and an incident safety officer?" Most of the firefighters can give me an answer, but most of the time they are wrong. There is a significant difference between these two positions, and we need to understand why.

To help us to understand, let's begin by looking at some terminology as stated by NFPA 1521, *Standard for Fire Department Safety Officer Professional Qualifications*.[1]

> **3.3.47 *Safety Officer.*** A generic title given to a member within a fire department or emergency service organization who performs the

functions of a health and safety officer, an incident safety officer, or who serves as an assistant to a person in either of those positions.

3.3.47.1* *Health and Safety Officer (HSO).* The individual assigned and authorized by the fire chief as the manager of the health and safety program.

3.3.47.1.1 *Assistant Health and Safety Officer.* The individual assigned and authorized by the AHJ [authority having jurisdiction] to assist the fire department HSO in the performance of the duties and responsibilities of the HSO.

3.3.47.2* *Incident Safety Officer (ISO).* A member of the command staff responsible for monitoring and assessing safety hazards or unsafe situations and for developing measures for ensuring personnel safety.

3.3.47.2.1 *Assistant Incident Safety Officer.* A member of the fire department appointed to respond or assigned at an incident scene by the IC to assist the ISO in the performance of the ISO functions.

As you can see by these definitions, there is a significant difference between an HSO and an ISO. However, in the past, the terms safety officer and incident safety officer were lumped together. As the use and requirement of the safety officer position became more a regular part of the fireground, we had to fill that spot. This usually meant we had to give up someone and many times, that person was a senior officer. Today, we need an ISO on every fireground. This position is an important position in any fire ground risk management system. So what is an ISO? First, the ISO is not a "gear cop." The issue of wearing gear properly, chin straps in use, hoods, and more, is the job of the company officer to enforce. The company officer needs to do his or her job, leading by example and by enforcement. The idea and concept of the gear cop came from members in the role of safety officer not knowing what the job really was or how to do it properly. So, they fell into the trap of familiarity, as firefighters were most comfortable with the doing part of the suppression activities on the fireground. As the safety officer, you may tend to watch what you know: the raising of a ladder, the chin strap, how the crew is lifting or lugging equipment, and other familiar acts. If they are not being done as you would have them, you consider it a safety issue and you intervene. I have even seen an old-time safety officer adjust the pattern on a handline being used from the exterior. *All of these issues that I just described are the company officer's job to correct.*

The job of the ISO today is to monitor the operational safety of those working at an emergency scene. As Battalion Chief David Dodson (retired) and author says, "Don't get caught up with the small stuff." The ISO is a key component of any incident action plan (IAP). Today, the ISO is a key position. At all scenes, the incident commander has overall authority for management of the incident and this includes all safety aspects. As an incident grows, designating an ISO relieves the IC of an additional job. The IC still retains overall responsibility for all aspects of the incident including safety, but the role and activity of operational safety is delegated to the ISO.

NFPA 1521 defines the role of the ISO as:

> **5.2.2*** *Monitor the IAP, conditions, activities, and operations, given an incident or planned event, an IAP, and risk management assessment criteria, so that activities and operations that involve an unacceptable level of risk can be altered, terminated, or suspended to protect members' health and safety.*

This is a very responsible position. Some years ago, my fire service mentor Assistant Chief Jack Peltier, Marlboro, MA, (now deceased) was beginning to teach me how to be a great ISO. He emphasized that the ISO needs to have the same or greater level of knowledge and skill as the IC in order to do his or her job. Jack taught me that the ISO is a position of authority *and anticipation*. The ISO needs to be monitoring the scene and constantly looking at what the companies are doing and what should be happening. The position requires one to be very proactive. You need to anticipate actions and look for and observe expected results. If the results are not happening, you need to advise the IC and suggest possible corrective actions. When you watch for expected results you also anticipate what might not go well and start to plan alternatives for success. You see, Jack taught me that a great ISO is a valued consultant to the IC, someone who can be trusted. The ISO does not tell an IC how to do his or her job, if a decision is wrong, or if the IC is making mistakes. The ISO advises, suggests, and points out observations and explains what those observations might mean.

A competent ISO constantly interacts with the IC, face-to-face preferably, with relevant information. In order for the ISO to do his or her job, the member must initially meet with the IC and get an overall situational status of what is going on, what the companies are doing, any known hazards and concerns, and how the plan is progressing (better or worse) upon arrival at the scene. He or she then needs to *conduct an independent initial size-up and 360 recon*. During the 360, the ISO should anticipate what should be happening. This size-up should be the first of many, especially if the incident is prolonged. (Remember, the longer it burns, the worse it gets.) After the ISO conducts an independent 360, the person meets with the IC and shares any observations that are relevant or important.

> **The ISO needs to be monitoring the scene and constantly looking at what the companies are doing and what should be happening.**

Back at the command post (CP) the ISO interacts with the IC, keeping abreast of the IC's plan, changes, what companies are exiting, what companies are committed, and what are their tactical assignments. The use of a command board greatly assists in this sharing of information (fig. 6–1). The ISO needs to know building construction, fire behavior, how to read smoke, how the fire will spread, and how the building might collapse. He or she must watch the building for stability, watch the smoke and fire situation, and watch the firefighters operating, which should be telling a lot about what is happening. Listen to the radio, it reveals a lot. Again, anticipate what should be happening based upon the tactics enacted. If what is expected is not occurring, something very bad could happen. This is the type of advice and counsel the IC is looking for.

As the incident continues, the ISO should be looking for any fireground hazards. This might include wires down, ice, leaking propane tanks, enclosed attack dogs that block access to areas of the fireground, and open ditches and trenches not initially observed. These issues are a danger to those operating on the

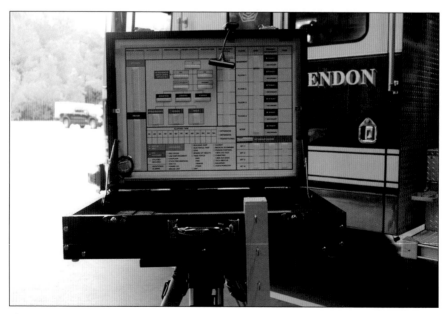

Fig. 6–1. A command board assists in the sharing of information.

fireground. These are just a few examples, but you are looking for anything and everything that might cause a firefighter harm.

The ISO should also make sure that a rehab area is established and used. Rehab is critical to firefighter safety and reducing risk. It is not the ISO's job to run rehab; the IC should assign a rehab supervisor. The ISO just needs to make sure it is put in place. In many parts of the country, rehab is automatically established and run by an EMS group or a community emergency response team (CERT). Regardless of who establishes and runs it, they are still part of the ICS system.

The ISO should make sure that the IC has appointed a rapid intervention team (RIT) and that the team is in place and capable. We discuss RIT in greater depth later in the next section.

The ISO is a key and critical position and not everyone can fulfill it. Different emergency scenes might require different sets of skills and knowledge. For example, a multi-vehicle crash scene typically has different hazards than a single-family home fire. As the incident commander, you need to make sure the ISO you select is highly qualified, even if it is a person of lesser rank. I have always said that the number of bugles on your helmet does not always mean you have all the knowledge needed at a given scene.

If you find that your officers do not meet the qualifications and requirements of a great and competent ISO, do something about it. There are a few excellent books available for the incident safety officer and, along with a copy of NFPA 1521, *Standard for Fire Department Safety Officer Professional Qualifications*, your department can develop a good program. In addition, if your state offers certification as an ISO, consider getting your senior staff certified. Finally, I would also strongly encourage you to require ISO qualifications as a part of any promotional process (appointed or elected) for all chief officers. Remember we are different!

The Rapid Intervention Team (RIT)

Every fireground should have a rapid intervention team (RIT). Now, you might call it a RIT, a FAST (firefighter assist and rescue team), or a RIC (rapid intervention crew). I could care less what you call it as long as you have one. So, what is a RIT? As defined by NFPA 1407, *Standard for Training Fire Service Rapid Intervention Crews*, 2015 edition,[2] it is:

> 3.8* *Rapid Intervention Crew (RIC). A dedicated crew of firefighters who are assigned for rapid deployment to rescue lost or trapped members. [1720, 2020] 3.3.31*

> *Standard for the Organization and Deployment of Fire Suppression Operations, Emergency Medical Operations, and Special Operations to the Public by Volunteer Fire Departments*

That is the NFPA "textbook" definition. My definition is this:

> *Rapid intervention crew (RIC, RIT, FAST). A group of firefighters, trained in both basic skills and RIT skills, who are well qualified and capable firefighters and who are standing by at the ready to make rapid entry, search for, and remove a lost, downed, or trapped firefighter before that firefighter dies or is consumed by the fire.*

To my knowledge you will not find rapid intervention listed in any of the formal risk management systems and books out there. Part of risk management is identifying and having the means to mitigate risk if it happens. A capable, staffed, and competent RIT should be part of each firefighter's toolkit to use if a firefighter is exposed to risk, gets in trouble, and has to be rescued. Remember the member is in trouble, with a very limited air supply, and needs help NOW! I strongly believe that staffing a capable RIT is a critical need in any risk management system. Here's why:

1. Let's start with the word Mayday. What does it mean to you? It is generally defined as "an international distress signal that means *help me.*" I have taken the liberty to introduce a different definition that I think is more in line with the use of the word in our fire service: "*help me or I'm probably going to die!*" Think about it: You are in an IDLH (immediate danger to life and health) environment, operating on a limited air supply which you are quickly exhausting, and something goes wrong. You are separated from your company, or you're lost or trapped in a partial collapse, or your SCBA has malfunctioned. You start calling Mayday! If able, you're looking for a window and exit point, or if you are deep within a structure, you're praying that they can find you quickly and bring you air to breathe! You are very much in a life or death situation. Having a RIT that is capable and well trained to answer your Mayday might mean the difference between living or getting a swell funeral.

2. As volunteers, we usually do not have a lot of extra help on our firegrounds. In fact, compared to our larger urban department brothers and sisters, we have a fraction of what they have. Some large departments respond on the first alarm with over forty firefighters first due. Large city FDs seem to always have at a minimum twenty-

five firefighters first due. I don't know about you, but where I live, we need to go to a third or fourth alarm to get that much help. When these cities strike a second or higher alarm they get ten to twenty (or more) firefighters right away. Now, imagine you're in a large department, have gone to a second alarm, and now have over forty-five firefighters working. If something goes wrong and you get a Mayday, you have a lot of qualified help on hand to deal with the problem. By striking another alarm or two, you will have additional help a lot faster than the typical volunteer organization. Now, let's have that Mayday come in on your volunteer fireground. How many people do you have initially and, if you called for more help, how many do you now have on scene? If you are a typical volunteer fire department, not too many. We don't have a lot of bodies to throw at the Mayday. Logic tells us that the people we do have need to be well trained and qualified to be on that RIT.

3. If a Mayday goes down on your fireground, there is no time to order in more help or to look at who you have on scene and who you can put on a RIT. We need that RIT to be moving right now, not in two, three, or five minutes. Please don't tell me that a few extra minutes won't matter. Imagine it's you, you have about five or six minutes of air left (I'm being generous), and when you call your Mayday, the IC has to look around to find some members to go and save you. You're going to be out of air and you have a high probability of dying before they find you. Would you like to wait those extra few minutes because your IC either does not believe in RIT, or gives the usual weak excuse of not enough people?

When I am out teaching or speaking to a group on rapid intervention and why a RIT is needed, I always ask four questions:

1. How confident are you of the firefighter survival and rescue training that your members have?
2. How confident are you that if the need arises, they can make a credible effort to save the life of one of our own?
3. Would you, without hesitation, put your own son or daughter's life in your RIT team's hands?
4. Are you sure?

When I ask the first two questions, I usually get a lot of head nodding and feel-good sounds from the audience. But, when I ask questions three and four, the audience tends to become very quiet. This dilemma is the underlying truth about RIT in many volunteer departments. We think that because of a few hours of training once, or the confidence that we never needed it before, means that logically, we will never need it in the future to protect our members. When you get a Mayday, is your member going to die or are you and your members going to give him/her as good a chance of survival as possible because you are trained, capable, and standing by?

Next, look at your members' RIT capabilities by analyzing the following four questions *honestly*:

1. What firefighter survival and rescue training has your department provided?
2. What survival skills have they been taught and how often do they drill on them?
3. Has the proper safety attitudes been instilled or are we acting reckless and out of control?
4. Are all the members trained in these skills, or only a few?

Let's look a bit closer at each one.

1. *What firefighter survival and rescue training has your department provided?* Have you conducted intense and not watered-down RIT training, or have you adapted it to fit the needs of your people? This usually is a sign that they are not capable. Proper initial RIT training is going to require 30+ hours at a minimum. This does not include the frequent refreshers needed on these unique skills. To help guide you in creating a true RIT training program there are textbooks available that explain the purpose and present the skills you need to learn step by step. A good example of this is *Fire Service Rapid Intervention Crews: Principles and Practice*. There can be no shortcuts; the training is hard and intense. Why? Because we are training to save one of our own, because our lives matter!
2. *What survival skills have they be taught and how often do they drill on them?* I have been conducting RIT training since 2000. One of the biggest jokes I see when I visit departments conducting training is to see the instructor demo a skill and then have everyone do it once. Really? How much can you retain doing it once? How do you develop motor memory skills doing it once? How do you retain the knowledge doing it once? I call this method of training the "one and done" training method. Training like this is a farce and it lets you fall into the complacency zone of feeling that all is well and you are content with what you and your members can do. Think about all that we have or do outside the fire service: football, golf, motorcycle

riding, hockey, soccer, you name it. None of these can be mastered by trying/doing it once and then you are an expert, so why do we think firefighter basic skills and RIT skills can be learned in one setting?

3. *Have the proper safety attitudes been instilled or are we acting reckless and out of control?* It was not that long ago I went to a fire as a spectator (in my area we call it sparking a fire) and watched an operation where the IC was clearly not in control. To assist in the lack of control, the firefighters were basically doing what they thought needed to be done. Breaking glass and venting without any coordination, applying water streams where they thought it needed to go, and other freelancing. From my point of view, it was reckless and out of control. When this happens, safety is affected. And, if the people performing in such an unprofessional way don't recognize or understand that safety is affected, we have a significant increase in risk, all because the department has not instilled the proper safety attitudes in their members. Safety begins with the proper attitude and mindset, and with the examples set by senior members. The "that's the way we do it here" attitude does not fly on today's fireground. With that attitude, you're asking for something to go terribly wrong. Safety is learning to take *calculated* risks, it is learning to do your job properly and with a common set of skills, it is learning to care about your life and the lives of the others on the fireground. It is not watching the movies *Backdraft* or *Ladder 49* and thinking that is what the job is all about, and it is not thinking that you know better than the person in charge and deciding to act on your own accord.

4. *Are all the members trained in these skills, or only a few?* Going with the "we are not all the same" theme I set early in this book, this is a tough one because we need to select those who are capable and focus the RIT training there. Why? Because in so many volunteer departments we have a mixture of young and old, physically capable and physically wanting. Fact, not fiction. So, if we look to all your firefighters who are trained and qualified (skill wise and physically able) to make entry into a burning building, these are the ones you need to make sure are all trained in RIT skills. By having them all trained and capable, you are not limited with who you can use for a RIT, nor are you scurrying looking for members.

The truthful answers to these four questions will assist you in pre-incident risk assessments and management. They will also assist you in identifying what actions you must take to staff a capable RIT every time.

Have you conducted intense, and not watered-down, RIT training, or have you changed it, usually by reducing the skill difficulty and physical challenges to meet what you think your members are only capable of? Why would you do this? Our lives matter! We need capable firefighters for RIT, not people with watered down skills who think they are great at it!

How we can staff a capable RIT?

When I talk with volunteers about RIT, the number one excuse given, every time, why they don't have one is, "We don't have enough people to dedicate four fire-fighters to RIT. I need them elsewhere!" I understand and have lived that, responding with three or four total and working hard until more help eventually arrived. As members of organizations that are typically much smaller than most urban departments, it is a fact of life that we respond many times with only a handful of firefighters. In rural areas, a handful might be pushing it. With that type of response, it is very difficult to dedicate four members as RIT. I understand this to some degree, but what I don't understand is where the lack of personnel excuse has been used to resist or avoid RIT training or staffing a RIT as soon as possible with the arrival of adequate personnel. Part of risk management is identifying and having the means to mitigate risk if it happens. A capable, staffed, and competent RIT is a tool (the means) to use for mitigation if a firefighter is exposed to risk, gets in trouble, and/or has to be rescued! Remember that the member is in trouble, *with a very limited air supply*, and needs help now!

So, I'm sure you are now saying, "So where do we get the people?" My answer is to start the process by embracing different ideas when it comes to rapid intervention if we want it available when it is needed. To begin with, it is reality that many (or you might argue most) of our small organizations cannot supply endless personnel for any incident. That is why we have mutual aid agreements with surrounding towns and other fire departments. It was and still is a good solution to provide more equipment and help. However, a mutual aid agreement does not always mean quality or capable help. I have seen mutual aid called for and the engine arrives with two people, one of whom is not allowed in a burning building. Now if all you needed was a pump operator or tanker, you are most likely all set, but if you were looking for a crew to make entry and assist in the suppression, the two who arrived are perhaps not very helpful to you right now.

The difference between mutual aid and mutual aid for RIT is significant. With RIT, we must have trained and capable personnel. to create and use a RIT mutual aid pact, we also must have a few other requirements. To begin with, we need consistency in the RIT skill sets taught and the training delivered. This means that every fire department in the mutual aid agreement must have the same RIT skill sets, with the same techniques and commands. To better understand this point, think of the basic skills (FF I/II) training your members receive. I am willing to bet that the vast majority of us have the opportunity to receive formal training in these skills, usually offered by state or county training groups. This is how we receive consistency in training. Because of that consistency in training, if/when a 35 ft ladder needs to be put into action, you can conceivability take four firefighters from four different departments, give the order, and the ladder will most likely go up quickly and efficiently. That is what common skill sets are supposed to achieve and why mutual aid works.

However, when it comes to RIT, things are different. If the mutual aid firefighters who arrive are not RIT trained, or have been trained with a different skill set, or are physically unable to fulfill a RIT activation, then you have a big problem. A mixed company cannot function together. If all your mutual aid departments have the same skill set and are dedicated to providing quality, capable (as in properly trained and physically able in RIT) help, you can build a RIT using members from different departments. Every member of every fire department that responds to mutual aid should be trained in the basics of RIT. If we cannot achieve this, then we are gambling and the *risk will increase* if a member gets in trouble and calls a Mayday. Why? Because there will be no one there with the skills to help rescue the downed firefighter.

Even with agreement in common skill sets, it does not address where does the extra help come from. I have pondered that question for years and can only come up with one solution. It is a different spin on mutual aid, but if you stop, accept change is needed, and accept we need to protect our people, it is the only answer. That answer is to create RIT mutual aid agreements. We all know on a county or region basis there are always *those* departments! You know who I'm talking about. The ones you never call mutual aid to the scene, and always put into station coverage if you even call them at all. The reasons are the same no matter where you live, and it all comes down to either lack of training, a reputation for recklessness, or incompetence. Most times these departments are led by people who live in the past, think progress is something to avoid, and take great pride in being maverick. We see these type departments as being reckless, afraid of change, poorly trained, and a danger to other firefighters. I once worked with such a person and it was very difficult for the members who wanted progress and change. We would watch equipment from other towns and departments respond by our station to major incidents and we were never called. It left so many of us angry and embarrassed. However, to paraphrase Voltaire, "It is dangerous to be right when those in power are wrong."

In a standard mutual aid agreement, it is assumed you will send capable people if called upon. In my proposed RIT mutual aid agreement, assumptions are not good enough. Build on a group of departments that have the same core training values as you. As you create this RIT group, come together and standardize the training and skills to be taught. Rapid intervention training is not a "do the skill, one and done." It needs to be ongoing. The skills need to be kept fresh (this means ongoing training) and the members staffing the crew need to be capable with knowledge, skills, and abilities (KSAs) and of high quality. Firefighters' lives depend on it. Encourage departments to share knowledge and expertise. Conduct joint RIT drills and strive to be the best.

I have firsthand experience in this type of venture. I participated to help create a group of firefighters who took great pride in the fact that they were the "go to" if things went bad and a firefighter needed to be rescued. They were the elite firefighters in their district. It was not just in title but in fact due to skills, abilities, and capabilities. The firegrounds they were on thus had a highly capable and ready RIT, not a group of four people thrown together to fulfill the standard who could not rescue themselves out of a cardboard box.

Once the training is complete, it is time for the next step: to get the RIT on scene quickly. How can you do this? By being proactive and calling a mutual aid dedicated RIT as an *automatic first alarm add-on to every structure fire with the initial alarm.* Yes this might be different (a change) from the way you have always done it, but remember change in the fire service always brings resistance. Some resisted SCBA, hoods, bunker pants, PASS devices, and many other improvements, but they were proven wrong. I recently saw a humorous post on Facebook: "There are two things firefighters hate: change and the way things are!" We need to move forward. There is no other place to get the help you need! I'm asking you to think and be different. Set up agreements with your surrounding mutual aid partners, get everyone trained the same way, and then automatically call them when you are toned out to a possible fire. Don't wait until you get to the scene to see what you have. If upon arrival have to wait 20 plus minutes plus for a RIT, you are placing those working on the scene at risk. If something goes wrong and there is no one there to rescue them, you will regret your decision for the rest of your life!

In closing out this discussion on rapid intervention, I would like to summarize it with these few thoughts: As volunteer firefighters, we provide a critical service to the community we serve. We and our communities are under a moral and ethical obligation to our firefighters to provide the RIT training needed to staff a group of capable, competent firefighters, standing by at the ready at every fire, to save the life of one of their own if called upon.

> **A real risk management program *begins* with your department training.**

Training as a Method for Risk Reduction

The following section on training is meant to be an introduction to bring awareness of how training can help to control and reduce risk, and also to share some thoughts on training. It is up to you and your department to take this information and expand and develop it into something that will be of benefit to all.

Why training matters

Our journey into managing risk is drawing closer to an end within the covers of this book. However, as was once said, "To make an end is to make a beginning"[3] and that is where I want to go next. If this book has you thinking about risk more than ever before, or if it has helped you to identify new methods and thoughts regarding reducing and managing risk, then you need to now go and make a new beginning. A beginning dedicated to a goal of reducing risk. This goal begins with training.

A real risk management program begins with your department's training. When things go wrong and someone is injured or killed, the training records are looked at along with the department's training program. Most times we can trace a direct link for what went wrong to a lack of or ineffective training.

Training is the *preparation stage* of everything in life we need to do. In our fire service it is the key to everything we need to have happen and hope to achieve at an emergency scene. Without proper training we are not prepared and are exposing ourselves to high risk, things going bad, and then making excuses for our poor performance. When I first joined we were seen as "only volunteers." The public did not expect a lot from us. The job had danger to it, but we were not as aggressive as we are today, nor were the fires as intense. Many times, upon arrival, we had a well- or fully-involved structure and the initial attacks were exterior or defensive. Today, the fires are much more intense, vent limited, contain higher levels of toxins, and are much more dangerous and hotter than ever before. This is a fact proven by recent UL and NIST studies.

Today, training is more of an issue in the volunteer fire service than ever before. A big part of that issue is that the work and family time demands on our members are significantly greater than in the past. In my early days there really was no Firefighter I/II curriculum. There were the basics of firefighting (firemanship) such as ground ladders, hose handling, SCBA donning, and how to use it. (By the way, SCBA was only used for firefighting then, not for overhaul, etc.) There were no exams, no certification, and no formal, structured training unless you were a member of a larger career department. We as volunteers were for the most part allowed to do things "our way." Many departments stepped up and made a strong training effort, and many others did not. Unfortunately, some departments still think and operate in this frame of mind, and that is dangerous and increases the risks we face. Today, it is very different. Many states have mandated levels of training required for all firefighters. These standards are typically based upon NFPA 1001, *Standard for Fire Fighter Professional Qualifications.*[4] It is important, even if your state does not have minimum qualifications, that your members receive training to the minimum level of Firefighter I/II. When I think of the word professional, I don't think of a career firefighter exclusively. I think of any firefighter who has the skills, knowledge, and abilities to do the job safely and effectively. To think that we as volunteers can lower our standards is a dangerous game to play and increases the dangers and risks your members face.

The bottom line here folks is that good training is the basis of everything else we need to keep safe.

What about your department's training?

Over my years of experience, I have developed some basic tenets of training that I would like to share with you. Training needs to be at least these six things:

- Punctual
- Safe
- Pertinent/applicable
- To the point
- As realistic as possible without compromising safety (fig. 6–2)
- Fun

Yes, fun! We are giving our time and it should be enjoyable.

When I am asked to discuss training with a group of chiefs I like to ask three questions:

1. If I walked in on your department's training night unannounced and unexpected, what would I see?
2. If I was to ask members at the training what they thought of it, and in general, of the department's overall training program, what would they tell me?
3. If I asked you what you thought of the department's training, and you had to be brutally honest, what would you tell me?

Let's explore these three questions.

Fig. 6–2. Put firefighters in realistic scenarios that are safe.
(Photo courtesy of Robert Gatchell)

1. What would I see if I walked in on your department's training night unannounced and unexpected? When I enter would I see people standing around acting bored, or would the firefighters be engaged in skill training?

   • Would the class have started on time? Today, we talk about our volunteers being time sensitive. That's fair. But it's not fair to have 6:30 PM drills start at 6:45, 7:00, or later waiting for members to show up. What about those who are there on time? What are you saying to them? Thanks for coming but you're not as important as the others I want to wait for? Drills need to begin on schedule. Don't waste people's time; show that you appreciate them attending training by always being punctual.

   • If it is skills training, is the instructor trained and capable to do the skills being taught? Don't laugh, I have actually seen instructors holding a book open looking at the pictures and telling the students what to do! Is the instructor physically able to demonstrate and do the skill?

   • Are there enough instructors to work with the size group you have? The optimum size for hands-on training is a 1:5 to 1:7 instructor to student ratio. This keeps everyone engaged, and the ratio allows for students to get more one-on-one time with the instructor to ensure they are learning the skills correctly and can execute them with competence.

- Were the equipment and props ready to go at the drill start time, or did we waste 20 to 30 minutes setting up? A good drill starts on time and that means everything (equipment, props, instructors, etc.) is ready to go. Everyone can help pick up and return everything to service as part of the actual drill time allowed. Again, people today are time sensitive. Show you appreciate and respect them being there and giving up their time by managing that time appropriately.

- Are the firefighters asked to do the skill once and then they are done, or are they required to complete each skill numerous times to build and show competence and muscle memory? Folks, one and done is not the way to train firefighters!

2. If I was to ask individual members at the training what they thought of it and, in general, the department's overall training program, what would they tell me?

- Do they feel the drill is relevant and educational, or is it the same old, same old? Things change. Have your training staff kept current and updated, or is it "this is the way we have always done things here" (fig. 6–3)?

- Do the firefighters feel the drill is a joke, not covering the material properly, or that the instructor is not demonstrating the drill properly?

- Do the firefighters tell me to look around and as I do, point out that it is always the same small group who attends the drills? In fact, I'm told the majority of members rarely show up as they have the attitude, "Been there, done that. I'm all set." My favorite experience with this was a hands-on skills class I was conducting for a volunteer department. It was open to all, but we only had twelve attendees. I asked the chief how many members were on the rolls and was told seventy-five. I then asked how many have gear and the answer was fifty, then asked how many go to calls regularly and the answer was twenty-five or so. Then I asked how many attend training regularly. The answer was, "the twelve you see here." Think about it, if you are one of the twelve, how do you feel going to calls with others who never attend training?

- Do they say to me, "Look at the officers attending." As I do, I observe if they are actively participating, leading the way, and showing their skills. Or, are these officers avoiding doing the drills, saying "I've done this a million times so I'm all set." I have

Fig. 6–3. Training exercises should be kept current and be relevant to what firefighters will likely face on the fireground. (Photo courtesy of Robert Gatchell)

always said that in our volunteer service, being an officer should not be a lifetime appointment. Company officers need to realize having knowledge in your head is one thing, but as a company officer you are expected to be with the crew, doing. If you cannot physically fulfill that role, why are you still a company officer? For many, this is difficult to hear and I understand. For me, when I got to the point I could not fulfill my role on the fireground, I retired from the department. I could have hung on, but that would not have been fair and in the safety interest of the members. Officers being able to do their job is a key component of any risk management system!

3. If I asked you what you thought of the department's training, and you had to be brutally honest, what would you tell me? I am asking you in the interest of safety to honestly assess your department's training program! Things are different now from perhaps when you started. Has your department's training kept up with those changes? Have you kept up? Everything changes: medical skills, automotive repair skills, computer skills, even how to operate our televisions today skills. So why are so many reluctant to keep up and change when it comes to firefighter skills? Our departments and our communities have an ethical and moral obligation to provide our volunteers with the best and most current training to keep them safe and protected.

What level of training is needed?

Generally speaking, in volunteer departments a variety of people and capabilities are represented. Past practices might have allowed a minimum of training that would in no way even come close to what is needed and expected today. The past is the past and frankly, good and safe departments don't live there anymore! Today we need to examine what skills our members need and then provide the training to meet those needs.

It is widely accepted that the minimum training standard is to meet the qualifications of NFPA 1001. Train your members who will be involved in structural interior attack to the level of Firefighter I and II. We are asking people to risk their lives responding to and entering burning structures. Personally, I also include on the list of must-have training the basics of rapid intervention. We discuss RIT earlier in this book and I remind you that unless the team has the proper training, who will be capable to save one of our own if the need arises? I've always been disappointed that basic RIT skills have not been included in Firefighter I and II skills. Unless our members understand the risks involved and have the skills to help avoid or limit those risks, they are in a high danger zone! There is no reasonable excuse to not train your members.

In addition to basic firefighter skills, they should also have training on driving and operating emergency vehicles. Depending on what type of calls you respond to, you also need to consider additional training. Let me share a great example of this. A few years ago at FDIC, I said hello to a firefighter I had met the year before at the show. We got into a great conversation about training. Now he was from an area in Illinois that is a major corn producer, and he started to talk about grain silo rescues. My immediate response was to ask, "What are silo rescues?" What I learned that day (besides what a silo rescue entailed) was that depending on where we are located, we all have unique type response incidents that we need special training on. What are yours, and do you provide that training?

In addition to the skills training, your teams need to learn department standard operating procedures and department policy. The federal government also requires all emergency responders who might respond to incidents that involve hazardous materials be trained to a HAZMAT awareness and operational level.

How much training?

This raises the question: Should training be mandatory? My initial response is: If you don't train, how do you learn and retain? Since the day I began, I have heard other firefighters complain about those who do not regularly attend training. I think that this "I'm all set, I know what I'm doing" attitude goes back to the old days, when the accepted practice was make the drills you can and don't worry about it. Today, we are faced with entirely different types of emergencies. The fires are a lot more complex and dangerous, the vehicles we cut open are a lot more challenging, the tools require different techniques than you might have used back in the day, hazardous material incidents are more common and very dangerous, the equipment we use is a lot more complex, and the fireground is run with a formal chain of command! And this is not a complete list! So with all these skill demands and requirements on us, why is training not mandatory for all members? NIOSH LODD reports show that the lack of training is a factor in many volunteer firefighter deaths. So again, why is training not mandatory? The people we protect and serve expect us to do the best we can to save them and their property. They don't care if you're a volunteer; they want you to do your job and the only way to do it is by having the proper training. I believe that every volunteer department needs to set a minimum training participation requirement for all its members. It might be 65%, 75%, or higher, but whatever you set as a minimum participation threshold, you need to stick to it. Remember, we want to be treated as the professionals we believe we are, and mandatory training participation will keep our skills sharpened and we will all be refreshed, renewed, and ready to respond.

> ## Why is training not mandatory?

Officer training and qualifications

One area we are lacking in is the training of our officers. In reality, most of the officer training programs offered today are much more applicable to the bigger jobs. One size does not fit all here. I remember in the early 2000s taking a strategies and tactics for the first due officer class. There were twenty-eight volunteer officers and only a couple career officers in the class. When it came time for the tabletop scenarios, every evolution gave us the same resources: four engines staffed with four members, two ladders (trucks) staffed with four, a heavy rescue staffed with six, and a district chief. That's thirty-one firefighters first due! What good is such training if it is not applicable to the needs of volunteer departments? In my opinion, it is not much good at all. So many of our volunteer organizations respond initially with three or four members. If you're a bigger organization maybe ten to fifteen. That day I learned how to be a company officer for a good size city and yet I was from a small town. Square peg, round hole; the training was very much inapplicable! We need to provide our volunteer officers with the training they need, whether they are elected or appointed.

Here are what I call my ten foundation stones of knowledge,[5] where I teach what all volunteer officers need to know to be able to do their job and why they need to know it.

1. How to be an effective leader
2. Fire behavior (including modern fire behavior)
3. How to read and interpret smoke
4. Decision making on the fireground
5. Understanding strategies, tactics, and tasks and incident action planning (IAP)
6. The brief initial report
7. Size-up—initial and ongoing
8. Building construction
9. Risk management
10. Accountability and crew integrity

Look at these ten subjects and think about what's so demanding or difficult about them if you want to be an officer and lead a company on the fireground. These are all officer basics. Every officer on every volunteer fire department needs to know and understand these ten subjects, regardless of the size of the department. If you enter or send firefighters into burning buildings, you need to know and understand this stuff. Please don't offer any excuses, because none will be accepted. Let's step up and get the classes that will train our officers so that they can operate safely and help to reduce or minimize the risks we all face at a fire!

Let's talk about the qualifications we require of our officers. Does your department have any? Most have a length of service qualification, but what it is varies. Please understand that length of service does not guarantee experience, capabilities, or knowledge. It just might mean a person has been a member for a certain period of time. In 2015, a 19-year-old lieutenant died in a structure fire.[6] In a previous chapter this LODD was discussed in greater detail. Trying to be as respectful as I can, I need to question how much experience did this person have in structural firefighting? You can't be a structural firefighter until you are 18 years old. I'm sure he was a very enthusiastic member, went to all the training, made a lot of calls, and was very active. But, what were his qualifications for being a company officer? His death is a sad tragedy, may he rest in peace.

One time I was discussing this very subject with an individual who was taking one of my classes. He told me that the first time he got elected an officer was because "it was his turn"! I didn't realize your turn was a skill as an officer needs to have.

Regardless of whether you elect or appoint, please develop minimum requirements that include specific training and skills. For example, to be a lieutenant I would suggest that you have been trained to the level of Firefighter I/II in accordance with NFPA 1001 and you must have training in the subjects I outline in my 10 foundation stones. I don't think this is asking a lot of the individual. After all, the people who will be entering burning buildings with this lieutenant as their leader and officer are putting their lives in the leader's hands! The levels of experience and knowledge need to increase for higher ranks.

Apparatus driver training

In Chapter 1 we discussed the LODD statistics as it relates to dying while in route. We need to now address how we can, in conjunction with slowing down, better teach our members how to be safer on the road. Does your department have a training program for all your driver/operators? If not, you need to have one, and everyone who drives the apparatus needs to participate in the class—no exceptions! The lack of driver training is cited in countless LODD reports and it is a key factor in many of the non-fatal accidents. Driving a car, dump truck, oil delivery truck, or any other vehicle is very different from driving an emergency vehicle. To start with, when driving a non-emergency vehicle such as a car or truck, you are not responding to an emergency scene. There is no adrenaline rush, no sense of urgency, and no sense of hoping to make the apparatus be first out the door. Part of any good driving program needs to be teaching that even though you are a responding firefighter in your *personal vehicle or apparatus*, you do not have the right to drive like demon possessed idiots. I know, because for so many years I was one of them. Then I got smart and realized how dangerous what I was doing was to myself, the other firefighters, and all the civilians on the road, and slowed down! This is a fact of life and we need to do something to stop such dangerous and unprofessional practices. There are several excellent driver programs available. Research them and bring one in. It is well worth it and it will help to reduce risk!

Any discussion on apparatus and driving always needs to always include the topic of mandatory use of seat belts. Does your department have this policy and, if so, is it enforced at all times, no excuses? If your department is lacking in this area, do you have the courage to enact a program to reduce the risk we are exposed to? Remember we are always reading about firefighters ejected from rigs involved in accidents. Why were they ejected? The answer is quite simple: no seat belts in use. None of you would ever let me take your child, grandchild, neice, or nephew for a ride in my car unless they were properly and safely seat belted in. Would you drive around with your infant without a proper car seat? No way! So, why do we allow this or look the other way when our firefighters fail to belt up? I think we need to get extreme with this. If I was your officer on the rig, it would not move until everyone was seat belted in, period. I would not care what the call was, how big the fire, or anything else. My first thought will always be your safety first.

Skills and the basics training

One of my fire service heroes is Lt. Pat Lynch, Chicago FD (retired). Pat once told me that we need to learn to execute our basics to the fullest. What this means to me is that everything we do is built upon our firefighting basics. If we don't keep our basic skills current and sharp, how can any of us feel that we are ready to do the job when called upon? So, let me ask you, what are your basic skills and the basic skills of your members like? When was the last time you threw a ladder, or dragged a charged hose line up to the second floor? Does your department conduct yearly refresher training on fire behavior and include the new science of fire? Any good fire department training program will include frequent drills that incorporate the basics.

At a loss of what to train on? Then open up the *Fire Engineering's Handbook for Firefighter I and II*, or Fundamentals or Essentials textbook and use it as a foundation for training. If you're doing a search drill, add in things like rescuing a victim, using the TIC, and SCBA emergencies. Be creative but keep it relevant! No one likes training that is unrealistic or outright dumb! How often do you take your apparatus drivers out and drill on what is expected of them at a call? Most anyone can go a lake or pond and draft water, but what about having them simulate pulling up to a scene, having a company (two members) stretching a line, and the pump operator charging the line from the tank and trying to establish a permanent water supply (hydrant or shuttle). There are a lot of components of this drill and that is what is going to be expected of that operator at a fire scene. Think about it, when was the last time you did something like this for a drill?

For so many, drills seem to have become boring and tiresome. It's the same old thing. Fifteen members standing in line, bored, waiting our turn, while one instructor runs the drill. What a waste of valuable training time! Multiple stations and multiple instructors will make the drills more educational and worthwhile and you will not be wasting valuable time. We as volunteers need to become more aware of the different type of members we have today and then make sure that our training methods and time requirements are realistic for all. For those of you who struggle with this, there are many ways of getting help. The National Volunteer Fire Council (NVFC) has an excellent textbook entitled *Volunteer Fire Service Culture: Essential Strategies for Success*.[7] It will help you in better understanding the people we need to be attracting today to our proud fire service. Things are different today, and we need to be aware of it in order to be successful in attracting new members and keeping all of our members as safe as possible. Another great resource for drill ideas is NFPA 1410, *Standard on Training for Initial Emergency Scene Operations*.[8] Annexes A and B have many great evolutions displayed and outlined that will help you to evaluate your department's capabilities and skill levels.

Summary

As members of the volunteer fire service we are all under an obligation to each other to make our emergency scenes as safe as possible. We need to work hard to discover and implement new methods of reducing or eliminating risk! This means we all must stay proficient in what we need to do (skills), constantly keep ourselves current (knowledge and skills), and be well versed in the changes that always seem to be happening. Finally, we must stay dedicated to always seeking out knowledge and learning to be the very best we can be!

Throughout my active on-the-line career and now as a professional fire service educator and teacher, I continue to hammer that message to everyone I stand before. If you who are reading this book are one of the experienced members, it is your job to nurture, council, and mentor our future leaders. They look to you for direction and example! Embrace and practice risk management, show them it is the right and correct way to conduct ourselves all the time. If you are a future leader, embrace this message, do the right thing, and always measure the risks before us. I leave you with this call to action: Train and train and train, and then when you think you have had enough training, train some more because firefighters' lives—and your life—depend upon it. Are you ready for that challenge? A lot of families are counting on you.

> **We are not "only volunteers" but rather firefighters who serve our communities in a professional manner, with professional attitudes and skills.**

Notes

1. Managing Risk: Individual Common Sense First Steps

1. Data compiled from NIOSH Fire Fighter Investigation and Prevention program, https://www.cdc.gov/niosh/fire/. See also David L. Hard, Suzanne M. Marsh, Timothy R. Merinar, Matt E. Bowyer, Stephen T. Miles, Murrey E. Loflin, and Paul H. Moore, "Summary of Recommendations from the National Institute for Occupational Safety and Health Fire Fighter Fatality Investigation and Prevention Program, 2006–2014," *Journal of Safety Research*, no. 68 (2019): 21–25, https://doi.org/10.1016/j.jsr.2018.10.013.
2. Hylton J.G. Haynes and Gary P. Stein, *U.S. Fire Department Profile—2015* (Quincy, MA: National Fire Protection Association, 2017).
3. Haynes, *U.S. Volunteer Firefighter Injuries 2012-2014.*
4. Haynes, *U.S. Volunteer Firefighter Injuries 2012-2014.*
5. *Volunteer Fire Service Culture: Essential Strategies for Success* (Greenbelt, MD: National Volunteer Fire Council, 2018).
6. Bureau of Justice Assistance, U.S. Department of Justice, "Public Safety Officer's Benefits Program," https://www.benefits.gov/benefit/1073.
7. Firefighter Cancer Support Network, "Taking Action Against Cancer in the Fire Service," 2013, https://firefightercancersupport.org/wp-content/uploads/2017/11/taking-action-against-cancer-in-the-fire-service-pdf.pdf.
8. 115th U.S. Congress, House of Representatives, *Firefighter Cancer Registry Act of 2018*, H.R. 931, https://www.congress.gov/bill/115th-congress/house-bill/931.

2. Identify and Understand Methods to Limit and Manage Risk

1. NFPA 1250, Recommended Practice in Fire and Emergency Service Organization Risk Management (Quincy, MA: National Fire Protection Agency, 2015).
2. NFPA 1250, section 3.3.22
3. NFPA 1250, section 3.3.23
4. NFPA 1250, section 3.3.25
5. International Association of Fire Chiefs and National Fire Protection Association, *Fundamentals of Fire Fighter Skills and Hazardous Materials Response*, 4th ed. (Jones and Bartlett, 2019), 177.
6. Paul Grimwood, "Survive the Flow Path," International Association of Fire and Rescue Services, March 29, 2018, https://www.ctif.org/news/survive-flow-path.
7. IAFC and NFPA, *Fundamentals of Fire Fighter Skills and Hazardous Materials Response*, 178.
8. Alan Brunacini, *Fire Command*, 2nd ed. (Phoenix: Heritage, 2002), 39.

3. Managing Risk

1. Data compiled from NIOSH Fire Fighter Investigation and Prevention program, https://www.cdc.gov/niosh/fire/.
2. National Fire Academy (NFA) of the United States Fire Administration (USFA), Managing Company Tactical Operations: Tactics (1999), http://fire.nv.gov/uploadedFiles/firenvgov/content/bureaus/FST/Complete-MCTO-T-StudentManual.pdf.
3. Christopher Flatley, "Flashover and Backdraft: A Primer," *Fire Engineering*, March 1, 2005, https://www.fireengineering.com/2005/03/01/272179/flashover-and-backdraft-a-primer-2/#gref.
4. Stephen Kerber, Study of the Effectiveness of Fire Service Vertical Ventilation and Suppression Tactics in Single Family Homes, UL Firefighter Safety Research Institute, 2013, https://ulfirefightersafety.org/docs/UL-FSRI-2010-DHS-Report_Comp.pdf.
5. David W. Dodson, "The Art of Reading Smoke," *Fire Engineering*, June 27, 2014, https://www.fireengineering.com/2014/06/27/236553/david-dodson-the-art-of-reading-smoke/#gref.
6. John Mittendorf and Dave Dodson, *The Art of Reading Buildings* (Tulsa, OK: Fire Engineering Books & Videos, 2015).
7. U.S. Department of Homeland Security, "Homeland Security Presidential Directive 5," February 8, 2003, https://www.dhs.gov/publication/homeland-security-presidential-directive-5.
8. Richard Ray, "Fireground Management: Prioritizing Tasks on the Fireground," *Fire Engineering*, March 30, 2015, https://www.fireengineering.com/2015/03/30/313462/prioritizing-tasks-on-the-fireground-p1/#gref.

4. Control, Reduce, and Eliminate Dangers

1. NFPA 1851, Standard on Selection, Care, and Maintenance of Protective Ensembles for Structural Fire Fighting and Proximity Fire Fighting (Quincy, MA: National Fire Protection Association, 2014).
2. Homer Robertson, "Simple Ways to Determine Your Fire Flow Requirements," *Fire Rescue*, February 1, 2010, https://firerescuemagazine.firefighternation.com/2010/02/01/simple-ways-to-determine-your-fire-flow-requirements/#gref.
3. NIOSH, Volunteer Fire Fighter Dies After Inhaling Super-Heated Gases at a Residential Structure Fire—New York, NIOSH report no. F2015-20, February 27, 2017, https://www.cdc.gov/niosh/fire/reports/face201520.html.
4. NIOSH, Volunteer Fire Fighter Dies After Inhaling Super-Heated Gases at a Residential Structure Fire—New York, https://www.cdc.gov/niosh/fire/reports/face201520.html.
5. NFPA 1901, Standard for Automotive Fire Apparatus (Quincy, MA: National Fire Protection Association, 2016).

6. Loren W. Christensen and Dave Grossman, On Combat: *The Psychology and Physiology of Deadly Conflict in War and Peace*, 2nd ed. (Millstadt, IL: Warrior Science Group, 2008).
7. Arthur J. Bachrach and Glen H. Egstrom, *Stress and Performance in Diving* (San Pedro, CA: Best Pub. Co., 1987).
8. NFPA 1001, Standard for Fire Fighter Professional Qualifications (Quincy, MA: National Fire Protection Association, 2019).
9. NIOSH, Volunteer Assistant Chief Killed and One Fire Fighter Injured by Roof Collapse in a Commercial Storage Building—Indiana, NIOSH report no. 2014-18, July 18 2018, https://www.cdc.gov/niosh/fire/pdfs/face201418.pdf.
10. "The NIOSH 5," Pass It On—Fire Training [blog], November 2, 2014, https://passitonfiretraining.wordpress.com/2014/11/02/the-niosh-5/.
11. "Governors Island Experiments," UL Firefighter Safety Research Institute, May 21, 2013, https://ulfirefightersafety.org/research-projects/governors-island-experiments.html.
12. See, for example, Paul Spurgeon, *Fire Service Hydraulics & Pump Operations*, 2nd ed. (Tulsa: Fire Engineering Books & Videos, 2017).
13. NFPA 1250, Recommended Practice in Fire and Emergency Services Organization Risk Management (Quincy, MA: National Fire Protection Association, 2015).

5. Managing Risk in the Volunteer Fire Department

1. Roger Lunt, "Communications Broke Down: An Excuse for a Serious Problem," *Fire Engineering*, June 7, 2016, https://www.fireengineering.com/2016/06/07/210437/communication-broke-down/#gref.

6. Additional Ways We as Volunteers Can Reduce Risk

1. NFPA 1521, Standard for Fire Department Safety Officer Professional Qualifications (Quincy, MA: National Fire Protection Agency, 2015).
2. NFPA 1407, Standard for Training Fire Service Rapid Intervention Crews (Quincy, MA: National Fire Protection Agency, 2015).
3. T. S. Eliot, *Little Gidding* (London: Faber and Faber, 1942).
4. NFPA 1001, Standard for Fire Fighter Professional Qualifications (Quincy, MA: National Fire Protection Agency, 2019).
5. Joseph Nedder, "10 Foundation Stones of Fireground Knowledge for Company Officers," *Fire Engineering*, April 25, 2014, https://www.fireengineering.com/articles/print/volume-167/issue-4/departments/volunteers-corner/10-foundation-stones-of-fireground-knowledge-for-company-officers.html.

6. NIOSH, Volunteer Fire Fighter Dies After Inhaling Super-Heated Gases at a Residential Structure Fire—New York, NIOSH report no. F2015-20, February 27, 2017, https://www.cdc.gov/niosh/fire/reports/face201520.html.

7. National Volunteer Fire Council, *Volunteer Fire Service Culture: Essential Strategies for Success* (Greenbelt, MD: National Volunteer Fire Council, 2018), https://www.nvfc.org/print-copies-now-available-volunteer-fire-service-culture-ess ential-strategies-for-success/.

8. NFPA 1410, Standard on Training for Initial Emergency Scene Operations (Quincy, MA: National Fire Protection Agency, 2015).

Index

SYMBOLS

15-40 Connection 18–20
360 degree 66, 114, 125, 131

A

accountability
    passport system 101–103
    system 100–101, 119, 126
    tag 101–103
action planning 40
adrenaline 12, 110, 149
aerial 83–84
age
    authority and 34
    experience vs. ability 107–108
"A Healthcare Provider's Guide to Firefighter
        Physicals" 17, 19
air
    oxygen-enriched 26
    supply 68, 134
airways 125
alarm 134, 140
alcohol 11
aluminum 54
aneurysm 9
annuity 10
apparatus
    cleaning 17, 19
    driver 149–150
    equipment and 76
    hydrant and 92
    organization 84
    safety 12–14, 89
The Art of Reading Buildings 63
"Art of Reading Smoke" 47
asphyxiation 9
assignment
    command board 103
    company 101–103
    search and rescue 63, 111, 134
    tactical 40, 99, 107, 114, 125
    task 40, 97
authority 34, 131

B

backdraft 26

balloon construction 53, 62
basement fire 64, 100
bidirectional flow 26
black fire 51
blitz 106
Brunacini, Alan 27
building construction
    books 63
    fire resistive 54–55
    heavy timber 58
    key observations 62
    lightweight 62, 121–122
    modern 25
    noncombustible 54–56
    officer knowledge 96, 100
    ordinary construction 56–57
    time burning 109
    types 53–54
    wood frame 59–62, 100
Burnet Mill fire 70–71
burn time, prolonged 109, 119, 121

C

cancer
    detection 20
    gear and 15–17
    incidence registry 20
    prevention 18–19
CAN (conditions, actions, needs) 116
capability
    company 27, 32–33
    on-scene crew 63, 111, 126
    personal 35
carbon monoxide (CO) 49, 83
carcinogens 15–17, 83
CDC 5
cellar hole save 123
CERT. *See* community emergency response
        team (CERT)
chain of command 147. *See also* command
        structure
chemical chain reaction 42
chin strap 77, 130
civilians, trapped or missing 63, 109–110
"Close Your Door" 47
clothes, high visibility 15
cocklofts 62

collapse
  floor 58
  risks 53
  roof 97
  structural damage and 121
  zone 62–63
command
  board 101, 132
  post (CP) 132
  structure 67, 126
common sense
  vs. emotion 110
  managing risk 1–11
  priorities 71
  responding 11–15
  tips 119
communications
  example 126
  fireground 64–65, 77, 80
  mutual aid 65
communities, small 109
community emergency response team
  (CERT) 133
company
  mixed 139
  officer 34, 64
  passport (CP) 102–103
complacency 2, 82, 136
complexity 2
concrete 54
conditions
  changing 30, 114–115, 140
  physical 117, 118
conduction 43
construction. *See* building construction
contamination 16–17
contractual transfer techniques 27
convection 43–44, 121, 122
coordinated fire attack 25
Corcoran, Jack 30
CP. *See* company: passport (CP); *See* command: post (CP)
crew
  integrity 96, 98, 104, 122
  leader 108
  short-handed 117, 122
cylinders 77

D

dangers
  control or eliminate 28, 75–109, 113
  fireground 132

identifying 28
indicators 120–127
understanding risk and 146
DC. *See* deputy chief (DC)
death. *See also* line of duty deaths (LODDs)
  civilians 110
  one-time benefit 10
  sudden cardiac 9
decision-making
  IC and 39
  officers and 96, 148
  situational awareness and 29
decontamination 17, 19
delegating 127
departments in U.S. 5
deputy chief (DC) 105
dispatching 109
Dodson, David 47, 63, 130–131
dogs, attack 132
downtown 57
drag rescue device (DRD) 15
drills. *See* training
driving 11–12, 14. *See also* road
drowning 9
dump sites 33, 69

E

education, public 46
egress 25, 82
embolism 9
emergency
  out-of-air 88
  responders 15
  risk 1
  types of 147
emergency scene
  accountability system 101
  communications 125
  ISO 133
  size-up 30
emotions 111
ensembles 77. *See also* turnout gear
equipment
  condition 5
  on scene 35
  proper 76–77, 113
  training 144
evaluation, ongoing
  indicators and warning signs 120–127
  questions to ask 114–120
  risk management 28
  severity 72

experience level 94
exposure
    cancer prevention 18–22
    victim 110

F

Facebook 140
facepiece 86, 89
Fahy, Rita 3, 6
falls 8
farms 54
FAST. *See* firefighter assist and search team
    (FAST)
fatigue 117, 122
FCSN. *See* Firefighter Cancer Support
    Network (FCSN)
FDIC. *See* Fire Department Instructors
    Conference (FDIC)
feedback from officers, examples 116
fill site 33
fire
    alarm panels 94
    cut on floor joists 57–58
    deep-seated 48, 49
    double in size 33
    exterior overlapping 54
    fighting in the past 25, 26, 120
    helmet 86
    load 99, 117
    location 31, 65
    responding 1
    situation 41–44
    suppression 4, 72
    tetrahedron 42–43
    twenty-first century 120
    vent limited 26, 46
fire behavior
    basic 38, 41–44
    flow-path 99
    four stages 43
    modern 38, 72, 141
Fire Department Instructors Conference
    (FDIC) 2
Fire Engineering 2, 27, 47, 60, 71, 150
firefighter assist and search team
    (FAST) 67–68, 88, 133
Firefighter Cancer Registry Act of 2018 20
Firefighter Cancer Support Network
    (FCSN) 17
firefighters
    career 5
    I/II level 95, 146

injuries 6
interior 33
number in US 5
volunteers 1
Firefighter Safety Research Institute 105
fireground
    communcations 64
    habits 76
    hazards 132
    management skills 3
    officers 130
    priorities 71, 110
    trackimg members 102–104
firemanship 141
Fire Service Rapid Intervention Crews:
    Principles and Practice 136
fitness 6
flashover 26, 44–45, 48, 51–52
floors, multiple 122
flow-path
    dangers 99
    definition 26
    identifying 45–46, 98
    warning signs 121
flow, water 80
four-wheel drive 11
fractures 8, 9
freelancing
    behavior 12
    safety and 137
    three-step program 115–116
    venting 47, 99
fuel
    modern load 25
    smoke 44, 48, 51

G

gallons per minute (gpm) 78–80, 98, 105,
    106–107
gas
    fuel 44
    meters 77, 83
gear. *See* turnout gear
"gear cop" 130
glass 54
government 5, 10
gpm. *See* gallons per minute (gpm)
grain silo 146
Grenfell Tower fire 54
grey smoke 51
growth of fire 33, 115
guideline vs. procedure 93

guts 119
gypsum 54

H

habits 76
Handbook for Firefighter I and II 150
handlines. *See* lines
HAZMAT 35, 146
heads-up display (HUD) 87
health
    age and 33
    cancer 20
    in the fire service 6–7
    insurance 10
    issues 117
heart attack 13
heat
    conditions 48, 110
    PPE and 117, 124
    turbulence and 49
helmet 15, 77
help 32, 122–123, 135
highway reponses 15
Homeland Security Presidential Directive 5 66
homeowners 73
hood 15, 17, 19, 77
hose company 65
hose/nozzle configurations 77, 78–80, 105
HSO. *See* officer: health and safety (HSO)
humidity 69
hydrant
    laying line 89
    SOG 91–93
    weather and 69
hydration 69, 118
hydraulics 105
hydrogen cyanide (HCN) 16, 83

I

IAP. *See* incident action plan (IAP)
IC. *See* incident commander (IC)
ice 69, 117
ICS. *See* incident command system (ICS)
IDLH (immediate danger life and health)
    environment 17, 19, 77, 87, 134
ignition point/temperature 49
implementation 40
IMS. *See* incident management system (IMS)
incident
    management system (IMS) 66–67
    safety officer (ISO) 129–133, 131

stabilization 72
incident action plan (IAP) 38–40
    adjusting with feedback 116
    command system 66
    ISO and 131
    officers 148
    ongoing evaluation 114
incident commander (IC)
    accountability system 105
    giving assignments 41
    IAP and 39
    ICS and 67
    ISO and 131
    role 23–24
    survivability assessment 111
    three step process before action 40
    updates from company 114, 116
    warning signs 124–127
incident command system (ICS)
    fireground control 119
    history and importance 66–67
    proper knowledge 97–98
    transferring command 111
income 10
information flow 29
initial report 96, 114
injuries
    burns 9
    by type of duty 6
    fireground 5, 71
    lack of skill or knowledge 119–120
insurance 10
integrity 104, 123. *See also* crew integrity
intellectualism 2, 23–24
internal trauma 8, 9
ISO. *See* incident safety officer (ISO)

K

Kessler, William 91
knockdown 118, 122
knowledge
    officers 96, 119
    proper 90–98
    skills and abilities (KSAs) 140, 151

L

ladders 77, 82–83, 130
laminar 49
law enforcement 14
lay, forward and reverse 89

leadership
  officers 95–96, 108, 148
  operational observations 76
  rank and 34
life safety 71–72, 110, 122
lightweight construction (LLW) 59–61, 100.
  See also building construction
line of duty deaths (LODDs) 6–9
  building construction and 53
  causes 4, 120, 124
  communications and 64, 135
  fire behavior and 41
  freelancing and 115
  human error 30
  IAP and 39–40
  ICS and 66
  lack of accountability 100, 123
  lack of experience 148
  lack of incident command 97
  NIOSH statistics 3
  preventable 7, 11
  report 64–65
  responding and returning 6, 13
  RIT and 67
  search and rescue 4
  by type of duty 7
  vehicle crashes 7–8
  water supply and 69
lines
  attack 25, 89
  hand 78, 130
  hose 4
  liquefied natural gas (LNG) 94
  supply 92
  what to pull 31, 48, 78
LLW. *See* lightweight construction (LLW)
LODDs. *See* line of duty deaths (LODDs)
London, England 54
loss
  family 10
  prevention 23
  property 71
lumber
  engineered 60, 100
  lightweight engineered 59. *See also* light-
    weight construction (LLW)
Lynch, Pat 150

M

management 24, 66
masonry 56, 58
Mayday

crew integrity and 104
RIT and 134–135
when to call 64, 81–82
medical problems 6
Mendon (MA) Fire Department 91
mentoring 151
mills 58
Mittendorf, John 63
motor memory. *See* muscle memory
"Multiple Company Tactical Operations"
  (MCTO) 39
Murphy's law 119
muscle memory 87–89, 136, 144
mutual aid
  agreements 139
  automatic response 83
  common terminology 67, 97
  equipment 35
  lack of training 123
  RIT and 138–139
  staffing level 32–33
  water flow 70

N

National Fire Academy (NFA) 39, 78
National Fire Protection Association (NFPA).
  See also NFPA
  LODDs 6
  risk control 24
  volunteer statistics 1, 3
National Incident Management System
  (NIMS) 66–67
National Institute for Occupational Safety and
  Health (NIOSH)
  decision flow chart 5
  reports 2
National Unified Goal (NUG) 15
National Volunteer Fire Council (NVFC) 7,
  18–19, 150
Navy 96
neutral plane 45–47
NFPA. *See also* National Fire Protection
  Association (NFPA)
  1001 94, 142, 149
  1142 78
  1250 23, 28, 108
  1407 133–134
  1410 78, 150
  1451 13
  1521 129–131
  1851 77
  1901 85

NIMS. *See* National Incident Management System (NIMS)
NIOSH. *See also* National Institute for Occupational Safety and Health (NIOSH)
  key recommendations 39, 41, 53, 64, 66, 67
  LODD report 85–89, 97
  vs. NFPA 5
noncombustible materials 54–55
non-fire emergencies 7
nozzle 79–80, 105. *See also* hose/nozzle configurations
NUG. *See* National Unified Goal (NUG)
nursing homes 94
NVFC. *See* National Volunteer Fire Council (NVFC)

O

off-gassing 48
officer
  assistant health and safety 130
  assistant incident safety 130
  fire truck 96
  "foundation stones of knowledge" 96, 148
  health and safety (HSO) 130
  incident safety (ISO) 130
  leaders 108
  minimum requirements 148
  qualifications 133, 148
  safety 129–130
  training 95–97, 145, 148
on-duty 7–9
"only volunteers" 2, 5, 94, 141, 151
operational observations 76, 109
operations chief 98
operators. *See* apparatus: driver
oriented strand board (OSB) 51
OSHA 38
out-of-date tactics 25
overexertion 6, 8
overhaul 73
oxygen 33

P

particulates 16
PASS device 86
passport system 101–103
Peltier, Jack 131
personal protective equipment (PPE)
  cancer and 16, 17, 19
  fatigue and 117

  proper use and cleaning 77–78
personnel. *See also* staffing level
  accountability report (PAR) 81
  warning signs 122–123
physical ability
  age and 33–34
  qualified 137
physical, annual 17, 19
pickups 14
plastics 25
PPE. *See* personal protective equipment (PPE)
predictable actions and consequences 25–27
preparation 141
pre-plans 70, 93–94, 109
pressure 49
priorities
  family 7–9
  incident 71–74
proactive 118
problem identification 40–41
procedure 93
progress reports 65, 116
property conservation 73–74
protocols 115
public relations
  calls for action 106
  homeowners 73
  stream 107
Public Safety Officers' Benefits Program (PSOB) 10
pump checks 84
punctual 143
"Put the wet stuff on the red stuff" 35, 42

R

radiation 43
radio
  calling for more help 32, 120
  coordinating water flow 92
  examples of useful feedback 116
  frequencies 65
  ISO and 132
  shorthand 125
  traffic 65–66, 80–81, 115, 117
  uses 64
  voice contact 123
rain 69
rank 34
rapid intervention crew (RIC). *See* rapid intervention team (RIT)
rapid intervention team (RIT)
  Firefighter I/II and 146

ISO and 133
staffing 138–140
terminology 68
training 87, 136
Ray, Richard 71
reassignment 107
rehab 114, 117–119, 133
rescue
chance of survival 111
risk evaluation 72
RIT and 108, 133
tactic assignment 63
Respondersafety.com 15
responding
common sense 11
LODDs 6–7, 13
safety 15
time 121–122
responsiblility 4
retirement 34, 145
reuses, building 59
RIC. See rapid intervention crew (RIC)
rigs. See also vehicle
gear 17
open top cabs 25
organization 84, 96
risk
"a lot to save a lot" 2, 28
calculated 72, 137
control 23–24, 27
crew integrity and 104
definition 4
ethical responsibility 24, 140, 145
family and 9–11, 135, 151
identification 37, 38
incident indicators 121–122
management 1–20, 140, 151
mitigation 138
mutual aid 139
personnel indicators 122–124
reduction 76–109, 129
tactical assignments and 107
today vs. past 3
training 141
unacceptable 131
risk assessment
community 1
definition 23
pre-incident 137
risk-benefit analysis 2, 27–28
risk management
additional methods 129

definition 23–24
four step system 28–29, 37, 71
IC and 27–28
individual first steps 9
RIT and 68, 134
scope of this book 1
system 113
risk transfer 107–108
RIT. See RIT (rapid intervention team)
road
rage 12
safety 12, 14–15, 149
rolls 144
run cards 32, 35, 70
rural areas. See also suburban areas
departments 5
response time 121–122
RIT 138
roads 15
staffing need 32
water supply 70

S

safety
first 149
ICS and 67
life 71
officer 129
propper attitude 136–137
salvage 73
SCBA. See self-contained breathing apparatus
(SCBA)
schedule 143
science, modern fire 42
search and rescue. See rescue
seat belts 13, 14, 149–150
self-assessment 90, 118
self-contained breathing apparatus (SCBA)
cancer and 17
NFPA report 85–89
risk reduction 77
training 141
"Shower within the Hour" 17, 19
silo rescue 146
simplicity 24
situational awareness
five questions 32–35
IC and 30
risk management system 28, 113
roadway 15
size-up and 114, 131

size-up
  initial questions 30
  ISO independent 131
  officers 96, 148
  ongoing evaluation 114, 125
  problem identification 40–41
skills, common
  basic training 150
  Firefighter I/II 94
  officers 119
  RIT and 134, 139–140
skyscrapers 54
small-town living 63
smoke. *See also* reading smoke
  color 51
  density 51–53
  four key attributes 48
  fuel-enriched 26, 51
  inhalation 9
  laminar 49
  reading 47–48
  survivability profile 110
  turbulent 49
  velocity 49–51
  volume 48
snow 11–12, 69
soot 17
SOPs. *See* standard operating procedures
  (SOPs)
span of control 66, 126–127
staffing level 32–33, 71, 122
standard operating guidelines (SOGs)
  formal example 91–93
  PPE 77
  simple example 91
standard operating procedures (SOPs) 13. *See
  also* standard operating guidelines
  (SOGs)
steel 54
stopwatch 109
strategy 40, 125
stress 6–7, 88
Stress and Performance in Diving 88
stroke 9
structure. *See also* building construction
  collapse 53–54
  damage 121
  engineered lumber 51
  fires 4, 104, 108, 140
  integrity 59
suburban areas 54, 70. *See also* rural areas
sunscreen 18
supervisor 98

survivability
  life safety and 72
  missing civilians and 63
  profile 109–111, 113
survival skills 136
survivors benefits 10
suspension 116
synthetic contents 25

T

tactics 40, 72, 125
tag system 101–103
"Taking Action Against Cancer in the Fire
  Service" 17
tankers
  driving 14
  equipment on scene 35
  extreme cold 69
  mutual aid 70
  response time 106
  water supply 89–90
taxes 73
teamwork 94, 104, 123
technology 145
temperature
  freezing 68–69, 117
  ignition 49
tenders. *See* tankers
thermal imaging camera (TIC) 77, 81–82
TIM. *See* Traffic Incident Management (TIM)
time, monitoring 99, 108, 119–120
tobacco products 17, 18, 19
toothpick construction (TPC) 59–62. *See
  also* building construction
towing 14
town meeting 10
TPC. *See* toothpick construction (TPC)
Traffic Incident Management (TIM) 15
training 142–150
  accountability system 101
  communications 80
  drafting 90
  driver 13, 149
  evaluating risk 72
  fill and dump 90
  handlines 79
  IC 95
  ICS 98
  instructor 143
  level Firefighter I/II 94–95, 139
  LODDs 7
  mandatory 146

membership turnout 34
muscle memory 88
radio use 65–66
relevant and realistic 150
repetitive skills 87
rescue 135–137
risk reduction 141–152
RIT 68
roadway 15
SCBA 87
search 111
survival 135–137
tabletop scenarios 147
TIC 82–83
warning signs 123
truss
bow string roof 53
lightweight 53
wood 59
turbulence 49
turnout gear
clothing 17, 19, 86
heat stress 123
shelf life 77
storage 16–17
suspension 116
washing 15
wearing properly 130

U

UAC. *See* universal air connection (UAC)
Underwriters Laboratories 45–46
universal air connection (UAC) 87
unreinforced masonry (URM) 56
urban areas 54, 121
URM. *See* unreinforced masonry (URM)
U.S. Fire Department Profile 2017 5
Uxbridge, MA 70

V

vehicle
crashes 7, 11, 13, 133
emergency 149
personal 12, 14, 149
traffic 92
velocity 49
ventilation. *See also* venting
coordinated 121

flow-path 26
horizontal 25
LODDs 4
venting
consequences 25
coordinating through IC 98–100
freelancing 47, 115
incident stabilization 72
vent limited 38
victim
condition of corpse 111
location 110
Voltaire 139
volunteer
fire departments 31
freelancers 116
health and wellness 7
injuries 5
LODDs 3, 7–8
obligation to safety 151
officers 147
percent of firefighters 1, 5
resources on hand 72
RIT 134
skill level 35
training 141
*Volunteer Fire Service Culture: Essential Strategies
for Success* 7, 150

W

warehouses 54
warning signs 120–127
water
damage 73
delivery rates 79
municipal system 89, 106
public expectation 106
relay pumping 117
supply 69–70, 89–93, 119
weather
effect on operation 68–69, 117
mutual aid 33
while responding 11–12
warning signs 123–124
white smoke 51

Y

YNK "you never know" syndrome 84